# Not the Same Tree

# Not the Same Tree

Angela Dyer

Pelem Press

First published in 2013 by Pelem Press, London
pelempress@gmail.com



A catalogue record for this book is available from the British Library.

ISBN 978 0 9573748 1 2

Book designed by Robert Updegraff
Printed and bound in the UK by Berforts Information Press

# Contents

**Alan Burgess, *Black Poplar, Household Mead*, 2010.***

*A fool sees not the same tree*
*that a wise man sees.*

William Blake

*Poems are made by fools like me,*
*But only God can make a tree.*

Joyce Kilmer

*God has cared for these trees,*
*saved them from drought, disease,*
*avalanches, and a thousand*
*tempests and floods. But he*
*cannot save them from fools.*

John Muir

# Introduction

Trees are, in the true sense, awesome. We human pygmies are dwarfed by their size, made juvenile by their age, overwhelmed by their majesty, dazzled by their diversity and the multiplicity of their uses. We depend upon trees. They are the lungs of the earth, its cleaners and filters. They are hosts to a myriad bugs and beasts, as well as to other plants; they provide shelter and shade, refuge and relaxation for animals and humans alike. They are used to support roofs, cover floors, fire furnaces, mature wine, make music and paper, fly flags and sails, cross the Atlantic and the Antarctic. We sit on them, eat off them, sleep on them, are born on, cradled in, and will probably be buried in them. They feed us with their fruits, just as they feed our imagination and our art.

Stories of trees, and especially forests, are woven through our folklore, myths and literature. Where would the wicked witch live but in a forest? Where are babes lost and foundlings found? The tree of life and the tree of knowledge are symbols in the world's religions; tree worship was once widespread, and the trees teemed with gods and goddesses. Christians do not forget that Jesus, a carpenter by trade, was crucified on a 'tree', a wooden cross.

Wood is astonishingly enduring. I recall the thrill of digging up a perfectly preserved woven wattle fence that had lain undisturbed for over two thousand years since it was made by the Romans. Two thousand? A water-well constructed of oak recently found near Leipzig may be seven thousand years old, thought to have been fashioned around 5000 BC by Neolithic men using stone axes. And England lays claim to one of the world's oldest known wooden artefacts, a spear made of yew believed to be 450,000 years old, discovered in Clacton-on-Sea.* But to put this in perspective, we are told that the evolution of trees began some 400 million years ago.*

However clever we have since become at inventing man-made materials, the strength combined with pliability of wood, its

**Cliché it may be, but sunlight through trees remains irresistible.**

versatility, not to mention its aesthetic qualities, are unmatched. The variations of grain and colour cannot be reproduced artificially, and attempts to do so remain artificial: no two pieces of wood are alike.

Not so long ago most of us wrote with a pencil, graphite encased in wood, or charcoal, itself burnt wood, on paper, crushed wood. We still write now, on our keyboards, of the family tree, of barking up the wrong tree, of failing to see the wood from the trees. We touch wood to ward off bad luck. Trees are part of our language, our history, our legacy. We too have trunks, are vertical and forked; like trees we often become warty in old age – 'knotty, knarry trees', in Chaucer's words.

This book is a celebration of trees. It is not out to bang a conservationist drum – there is a constant drumming, whether we take note or not, by people better qualified than I – but to open the eyes of the ordinary person to the extraordinary importance of trees, on a personal as well as a global level. Nor is it scientific or botanical; there are many fine

**A knotty, knarry oak in the ancient hunting forest of Hatfield.**

**Caspar David Friedrich, *Wood in Late Autumn*, 1835.**

books under both headings. It is more a butterfly's flit through orchard, park and woodland, at home and further afield, alighting on whatever catches the eye: the magic of trees themselves, man's relationship with them and the part they play in our lives, our use of them in both practical and artistic terms, and our interdependence in an age when trees can no longer be left to do their own thing, to spread and sprawl and breathe as they have done for so many millions of years. For if throughout our short history we have needed trees, and still need them, they also now need us.

## PROLOGUE

# *Trees in the past*

That the whole of Britain was once covered by trees is probably not news to many people; it is one of those facts that we learn at school and promptly forget, whether we live in a city or not. But I was not prepared to discover that 'in the highlands of the central Sahara beyond the Libyan desert . . . the mountaintops, today bare rock, were covered at this period with forests of oak and walnut, lime, elder and elm.' The period referred to was a very long time ago, two thousand years ending in 3900BC – a time when large swathes of the Middle East were covered in cedar forest. Deforestation is not new: the story of the destruction of trees in this area is told in the *Epic of Gilgamesh*, arguably the oldest known written story in the world.*

One may think of trees that thrive in desert conditions as being hardy types such as acacias, figs and date palms, but in ancient Egypt willow and sycamore were common. In Egyptian mythology, the two 'sycamores of turquoise' standing at the eastern gate of heaven were associated with the goddesses Nut, Hathor and Isis, known as Ladies of the Sycamore. Willow was sacred to Osiris, son of Nut and wife of Isis; a grove of willow was said to have sprung up around his coffin, a symbol of the cycle of death and new life.

Opposite: **Ancient oaks in Wistman's Wood on Dartmoor.**

Right: **A tree with arms and a crude breast offers sustenance to the deceased.**

**A very long taproot enables the kiawe tree, a relative of the acacia, to survive in one of the driest deserts in the world, near Ica in southern Peru.**

In contrast to the aridity of the desert, the damp heat of the tropics provides conditions in which trees flourish to such an extent that they can reach the unimaginable height of over a hundred metres. Below these emergents lies the dense canopy of trees that shuts out 95 per cent of the sunlight, leaving the understorey and forest floor seething with plants and wildlife; it is claimed that half the world's species of flora and fauna are found here.

Though the forests of Britain are less prolific they do support a vast array of plants and animals, in systems both symbiotic and exploitative. From mammals such as deer, badgers and foxes, to the many different rodents, to a host of birds and insects, all rely on the cover of woodland and the individual trees for their livelihood and protection.

And so, throughout history, has man. Quite apart from being a vital food supply, with its nuts and berries and the associated fungi, and a source of shelter and firewood, woodland has long kept people in work. Foresters, woodcutters and carters; sawmill workers, builders and carpenters; charcoal burners, coppicers, hedge layers, makers of fences and baskets – where would all these have been without woodland? Today the cutting may be with chainsaws, the transportation by crane and lorry, the wood 'engineered' before being used, but the supply of wood is just as necessary as it has ever been, and provides as many people with work.

People have found uses for different types of wood since very early times. We may imagine primitive craftsmen making simple hollowed out forms as basic tools and implements – which they did – but extraordinarily intricate and sophisticated pieces of furniture, carved and elaborately inlaid in juniper, maple, walnut, box and many other woods, dating from the early first millennium BC have been discovered at the ancient city of Gordion,* in what is now Turkey. At this time the Egyptians were already turning wood on a lathe to make bowls and other artefacts, in 28 different types of wood only four of which were common in Egypt, the rest being imported. And from the ruins of Herculaneum, preserved in the ash that had buried it in the first century AD, came a wooden cot, still movable on its carbonised rockers.

In Britain, wooden supports dating back to 250BC have been found in a series of jetties at Poole Harbour. Since then the use of wood has been a constant, both as fuel (indirectly as a source of fossil fuels and directly as firewood and charcoal) and for building and implements. Hazel for hurdles, willow for baskets, alder for clogs, birch for brooms, beech for chairs, oak for almost anything – the list is inexhaustible, and all these are woods that come from our native British trees, never mind the naturalised species and more recent exotic imports.

The subject of native trees is disputed, and has differing criteria, but without getting too deeply into the arguments we can safely claim at least thirty species that have been in Britain since the last ice age ended 10,000 years ago. Any trees that came after that are classified as 'naturalised'. Many of the trees under the heading of native were introduced by man; for example the strawberry tree, a surprising native since it likes hot sun and flourishes in the Mediterranean, was probably brought from Iberia to southwest Ireland where, known as the Killarney strawberry tree, it survives in the wild today, unique in the British Isles.

The only trees with claims to be endemic to Britain, that is to have arrived without the help of man and before the English Channel cut us off from Europe, are some whitebeams, members of the Sorbus family. All of these could be labelled Celtic, as they occur in the west of England, Wales, Scotland and Ireland, and are very localised.

**A Scots pine in its native setting, below Ben Nevis in the Scottish Highlands.**

One of the earliest native trees was the Scots pine. This magnificent tree, growing up to 45 metres yet always graceful, is extraordinarily adaptable and tolerant, flourishing and providing its valuable timber – a softwood widely used for furniture – in very different climates, from Spain to Scotland to Siberia.

Some of those claimed as native are borderline as trees – low-lying shrubby plants, often with more than one stem, such as juniper, dogwood and buckthorn which seldom struggle to a height of more than eight metres. Useful they are though: from the dogwood were made the 'dogs', butchers' skewers that give the tree its name; juniper was used for making longbows, and its complex knot patterns make it a valued wood for marquetry; even buckthorn, not much use for its wood, was sought for the purgative properties of its berries – and those who couldn't wait for them used the inner bark, with sometimes disastrous results.

***Rhamnus cathartica*, purging buckthorn.**

The term 'ancient woodland' is used to describe woods that have been in existence since 1600 or before: a blink of the eye compared to the timescales we have considered. These woods were natural, that is not planted – though often managed – by man, and were therefore diverse in their species. In the category of ancient woodland were the medieval royal forests in which the hunting, mainly of deer, was a royal prerogative, with severe punishment for trespassers and poachers. Kings had forests (though the term 'royal forest' and the restrictions that applied to it covered grassland and heath as well) while commoners, if they were lucky, had access to woodland for firewood and for grazing their animals, and others harvested the timber for charcoal, housing and shipbuilding and later as fuel for the iron foundries and potteries.

**The Tree of Life as symbol takes many forms in religions and other belief systems; we may hang on it whatever we choose.**

Poets, writers, artists, musicians have all celebrated the tree throughout history – quite apart from relying on it as pigment, writing implement, paper and instrument. Trees play a prominent part in mythology all over the world: the gods and goddesses of the ancient Romans and Greeks, the tree spirits of the Celts and Druids, the tree of knowledge in Christianity, the tree of Bhodi in Buddhism, the tree of life in many religions. One of England's best known legends is that of Robin Hood, elusive green figure of the forest (which forest is a source of dispute, though it matters little to the story) whose exploits have captured the imagination of storytellers and writers from medieval times to the present day.

Many trees have associations with particular countries, some of which are obvious, others less so. In fact one could claim that some have national characteristics. The English link with the oak is longstanding, from King Alfred planting one to Charles II hiding in one and Tennyson writing a poem about one. *Quercus robur*, meaning strong oak, grows all over Europe and east to the Caucasus as well as in north Africa, but has been adopted – one might say appropriated – by the English, who call it the English Oak and treat it as a national emblem. The British do after all have 'hearts of oak', a phrase taken from the *Aeneid* and used to praise the strength of naval sailing ships; the Royal Navy still marches to the thumping words, 'Hearts of oak are our ships, hearts of oak are our men'. The oak, a fitting symbol of the English: solid, dependable, not easily moved.

Germany and the United States also regard the oak as their national tree, but theirs are just *Quercus*, not the particular 'strong' oak of the English. Other national tree emblems are most notably the maple for Canada and the cedar for Lebanon, which feature on their flags, the cherry for Japan, the olive for Greece, and for China the ginkgo. Trees have their place in a nation's culture, and 'a culture is no better than its woods', as W.H. Auden observed.

# Trunk

It may seem perverse to start in the middle, but the trunk is the essential ingredient in a tree's anatomy, without which the rest of it cannot exist. In comparison to the intricacy and variety of roots, branches, leaves, flowers, the trunk may seem at first sight a necessary but rather dull part of the tree's make-up: its support and prop, mere provider of sustenance to the more glamorous extremities. But take another look. Its surface may be smooth, flaky, furrowed, knobbly or twisted into a spiral; it may be silver, or yellow, or black, or grey, and only occasionally drab brown; it may be as solid as oak or as whippy as willow, as upright as a young poplar or as bent as an ancient hawthorn. Beneath its solidity the trunk seethes with energy, a mass of whirring particles and fizzing minerals as food and water are hoisted up and down in the sap to feed the demands of its dependants. (The word 'trunk' too meaning a duct or vessel, as well as the large box originally hollowed out from a tree.)

The trunk can enter the ground as cleanly as a metal bore, or it may reach out into a fan of roots on the surface of its chosen site. When a bank collapses one can sometimes glimpse the extent and energy of the root system, and from this understand how the trunk, in its supporting role, is itself supported: the central shaft bearing the mirror image of roots and branches, the axis linking heaven and earth.

Much of the tree's character, its form and bearing, derives from the trunk. When I think of my special childhood tree, the beech, it is the trunk that springs first to mind. Beech trunks are grey and very smooth, an invitation to amateur engravers of initials and entwined hearts. For me, the trunk typifies the gravity of this beautiful tree – which in my imagination was always masculine, though in folk lore it is known as 'mother of the woods' for its protective qualities. But my protector was unequivocally male, borne out by the Bach flower remedies which prescribe beech for the stereotypically male

**An old sweet chestnut spills its innards, but what remains will regenerate.**

**Mature beeches lining a bank, Draycott Sleights, Dorset.**

characteristics of pride, arrogance and over-strong will. In my childhood games it was the trunk that mattered – or rather, that place where the trunk begins its deployment into roots, which in the beech provides a landscape of deep valleys, paths and hillsides in which I designed a miniature world constructed of twigs and moss. European legend ascribes to the beech ancient wisdom and learning, and the word 'beech', through the Anglo Saxon *boc*, became book. (In Swedish, *bok* is both book and beech.) Certainly, like Saint Bernard, I learnt from my beeches things that I would otherwise not have understood.*

Most people it seems have a favourite or special tree, and most of these links are formed, even if subliminally, in childhood. And what a choice we have. Considering only the trunk, think of the wealth of subtle differences between that smooth silver-grey beech and others of our native trees, let alone the established immigrants. Of the giants there is the oak, whose dark trunk is rough and deeply furrowed, given to bulbous warts; the ash, at first sight similar but whose furrows have a diamond-shaped pattern; the elm, with bark of a blueish tinge and intersecting ridges; the lime, its shallower ridges encouraging the moss to cling; the deeply corrugated orange-brown of the yew. Then look at the shapes: that lime trunk, slim and elegant in its youth but becoming gnarled and lumpy with age; the convoluted hawthorn, with its tangled confusion of spiky branches; delicate, sinuous birches, stumpy pollarded willows, conifers with firm, straight trunks heading for the sky. How can one possibly *have* a favourite?

The human body has been likened to the tree in art and myth – after all, we share the name for our torso. The Greek naiad Daphne, turned into a tree as she tried to escape from Apollo, is captured in one of the world's masterpieces of carving by the seventeenth-century Italian sculptor, Bernini. That this is carved, and in marble, is almost beyond belief: the transformation from human body to tree seems to be taking place before one's eyes, like a slowed-down movie frame. A different transformation, hatched in the mind of the artist, can be seen in one of Giuseppe Arcimboldi's well-known vegetative caricatures, *Winter*, in which the wizened face emerges convincingly from the trunk-like neck. The contemporary sculptor David Nash creates human figures by firing tree trunks on a massive scale, like this majestic piece set amongst the living trees in Yorkshire Sculpture Park.

**David Nash, *King and Queen*, 2008.**

Classical carvings of standing human figures, lacking modern methods of reinforcement, were often posed against what Henry Moore described as a 'silly tree trunk' for support. Moore himself preferred to carve horizontal figures, and one of his key early works was a reclining woman carved in elmwood.* Many years later, at an age when he too had begun to feel the effects of arthritis, he drew the severely twisted hands of the nuclear physicist Dorothy Hodgkin, and shortly after turned to a series of studies of the tangled old trees in the hedgerows of his Hertfordshire home. One of these, subtitled *Knuckled Trunk*, bears a strong resemblance to the knuckled hands; other drawings were subtitled *Gnarled Roots* and *Tortured Trunks*, and looking at these I began to regard the convoluted twinings of my own varicosed legs as badges of honour, signs of hard knocks and stormy seasons survived, rather than something to be ashamed of. Trees, as Michael Viney observes, are honest about their age.

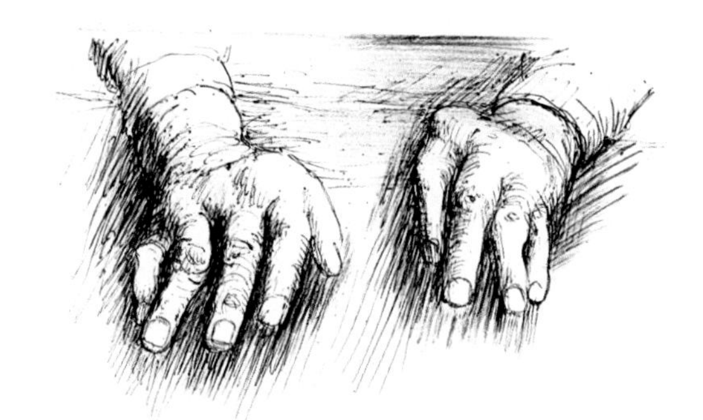

**Henry Moore, *Hands of Dorothy Crowfoot Hodgkin II*, lithograph, 1978.**

**Henry Moore, *Trees III: Knuckled Trunk*, etching, 1979.**

And astounding ages they can be – up to 5000 years old is the claim for some bristlecone pines in California, and Scotland boasts the oldest tree in Europe, the Fortingall Yew, judged to be well over 2000 years. Yew trunks are particularly gnarled, and what appears as a single tree may in fact be composed of many trunks. Yews, which grow very slowly, often die off from the middle, with internal shoots then taking root in the hollow in a cycle of potentially everlasting regeneration. This reputation as the eternal tree of life has, conversely, given the yew its common links with death. The girth of the Fortingall Yew was given as 16 metres when it was first recorded in 1597 – much bigger than it is now – but this pales into insignificance compared to claims for living giants such as a sweet chestnut of 60 metres and a baobab of 43 metres in circumference, though these are most likely composed of several trunks fused together.

The science of dating trees by their rings, dendrochronology, is precise and has many applications. Cross-sections of tree trunks are familiar to us all from felled trees and from schoolroom diagrams, even if their functions remain obscure. Put very simply, at the centre of the tree lies the aptly named heartwood which though dead supplies the tree with its structural strength, and around this are formed the annual growth rings, source of information not only about the tree itself – good years and bad, drought and plenty, the effects of fungal and animal attack – but also supplying data for other disciplines: climatology, archaeology, architecture, art history and, more recently, radiocarbon dating.

*Trees are poems that earth writes upon the sky.*
*We fell them down and turn them into paper,*
*That we may record our emptiness.*

Kahlil Gibran*

The felling of trees is a traumatic event, whether through natural forces or by the hand of man. The creak, the shudder, the slow sway, the crack and the astonishing force of a big tree falling make us feel, as indeed we are, powerless in its face. I don't subscribe to the belief that trees are sentient, but it would be unnatural to watch this and remain unmoved. We must feel it, even if the tree does not.

A tree on the ground is shocking, as shocking as the body of an elephant, say, in its bulk and immobility. I vividly recall walking down Rosebery Avenue in London on the morning after the great storm of 1987 – there were no buses – and being unable to grasp how so much tree, such profusion of branches, such immensity of trunk, could suddenly have appeared where previously the line of upright plane trees had been scarcely noticeable. As when we see a human body on the ground, the sudden horizontality of that which is normally vertical foreshadows our own return to the earth.

Felling is necessary of course, for despite man-made materials and synthetics we still use wood in massive quantities. Trees become wood – or lumber or timber, equally lifeless words – as soon as they are felled, in much the same way as the pig becomes pork: in losing life they become no more than objects to be used. But vast areas of woodland and forest are cleared not primarily for their timber but for the land, and often the trees are just destroyed.

We feel outraged when trees are needlessly cut, whether in the rain-forest or nearer home – and whether the 'needlessly' is disputed or not. David Hockney's late return to his native Yorkshire unleashed a flood of creative energy. His studies of trees and landscape throughout the seasons opened up a new world both for him and for his admirers. In the summer of 2008 he painted a stand of beeches and sycamores near the village of Warter in the Yorkshire Wolds, returning that winter to paint the trees in their outlines; but when he came back the following spring, all that remained was a pile of logs. He said: 'To me even the approach to that little wood had a kind of grandeur, like the approach to some marvellous great temple, and the trees themselves were very large, very architectural, very majestic. . . . It was like coming into some little village or town and finding that overnight the people had obliterated a great church that had stood there for 900 years.' Hockney assumed that the trees had been cut for their timber, but the real explanation was much more bleak. As reported, 'The copse fell foul of dreaded health and safety concerns. It was deemed to be too close to nearby cottages, and was cut down to protect them.'

The subject of wood is boundless. It comes in such variety, of texture, density, strength, colour. It has so many uses. It can be sliced, sanded, polished, varnished, painted. It must be treated with care, dried and seasoned in the right temperature and humidity. Because wood, though dead, lives on.

Rather obviously, the main source of sizeable wood is the trunk. Boughs and smaller branches have their own particular uses, but it is the trunk that supplies wood in the quantity and dimensions that have given this supremely versatile material its key role in our buildings and our furniture down the ages; now that we have grown clever in joining sheet wood, and using offcuts and waste to provide a sort of synthetic wood, the size of the trunk is no longer so important.

Of our native trees, the most desirable for its wood, historically and to this day, is the oak. Its durability, allied to great strength combined with flexibility, made it invaluable in shipbuilding and construction as well as for its many domestic uses. Beech is also heavy and strong but less resistant to the elements and also prone to attack by insects, so it was used for a variety of small implements and utensils before the advent of man-made materials, and still provides us with such things as clothespegs, handles and bentwood chairs. Being very durable in water, elm was used for boatbuilding and the construction of docks and jetties, for weather-boarding houses and also for kitchen tools. These three stalwarts all produced good charcoal, elm charcoal also having medicinal applications. Distinguished for *not* burning was poplar wood, which because of this was put to use as flooring in kilns and oast houses, and is still used for making matches.

Several of the trees we may consider primarily as fruit trees are prized for their wood: the apple, pear, cherry and mulberry are all sought after by carpenters, cabinetmakers and carvers, the warm tones of cherrywood being particularly effective in marquetry. And applewood smoke gives its distinctive flavour to meat and cheese.

Exotic woods are now available anywhere in the world and are much used for utilitarian furniture as well as for finer work. In *The Forest Stations* William Fairbank lists the 139 different woods that make up

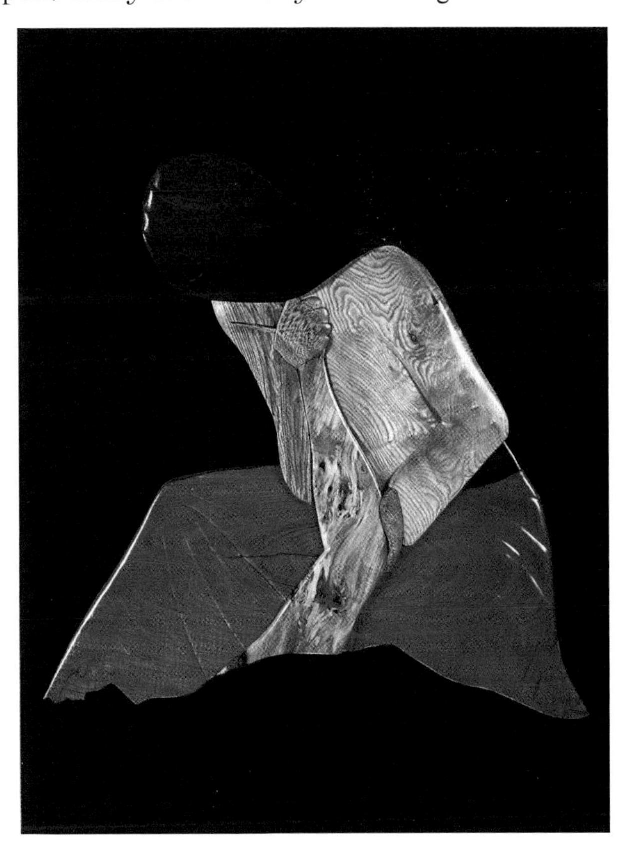

***The Forest Stations*, 1996. Mary's skirt is made of cherrywood, her torso of ash, and Jesus's arm is of lime.**

his sculptures depicting the Stations of the Cross, woods with delightful names such as bubinga, grendillo, ziricote, padauk, tsipaty and a host of tongue-twisters from New Zealand. Many of our common naturalised trees – the pines, spruces and larches, sweet and horse chestnut, sycamore – provide useful woods, as do the lesser known species made more familiar during the spate of landscape gardening in the eighteenth and nineteenth centuries, such as the cedar of Lebanon, the tulip tree, and the range of exotic maples, flamboyant cousins to our humble native field maple.

Going back two thousand years to the Roman occupation of Britain, the wooden writing tablets excavated near the Roman Wall at Vindolanda, the thickness of cardboard and folded in two like the leaves of a tiny book, were made of local woods such as alder, birch and oak. And the first books made of paper were encased in thin sheets of wood, usually beechwood, inside leather covers.

Wood engraving was much used in the fifteenth and sixteenth centuries, but the master of engraving was Thomas Bewick, who in the eighteenth century revolutionised the technique by working the endgrain, that is cross-section, of the wood – mostly very hard boxwood – and working it with the fine tools of the metal engravers. This made possible the illustrated book, Bewick's speciality; his histories of quadrupeds and British birds with their detailed and realistic engravings are legendary, and although he did not specialise in depicting trees, they form a constant backdrop to his subjects.

Musical instruments have been made from natural materials since the first flute was plucked from the reed bed. Wood was, and remains, the chosen material for wind, string, keyboard and many percussion instruments. 'Orpheus with his lute made trees' wrote Shakespeare,* but the trees were there first. Instrument making is a highly skilled art, and well-made pieces are immensely valuable and become more so with age; a Stradivarius violin made at the end of the seventeenth century sold for $16 million in 2011. Many types of wood are used, as solid woods and veneers, often in the same instrument and according to physical factors of weight, density, strength and flexibility, aesthetic ones of colour and grain, as well as the more elusive – and disputed – tonal qualities. Stradivari used spruce, willow and maple among other woods, including lime which came to be known as basswood in America. It has been suggested that the superior quality of his violins was partly attributable to lower than average temperatures in the previous half century, which resulted in slower growth and therefore added density of the wood he used, especially spruce. A musical instrument is the prime example of wood remaining 'alive', almost having a personality, as testified by musicians' attachment to their instruments.

Not all the trees felled end up as wood in its original form. Sawdust, wood chips, laminates, veneers and the ubiquitous MDF, medium-density fibreboard, are an integral part of modern building and furnishing. Some veneers are so thin that it is impossible to imagine how many layers might be shaved from one hefty tree trunk. But this economical use of wood, and of what would otherwise be waste, makes good ecological sense.

Imagining the world without paper is almost as hard as imagining it without trees, and wood pulp is the main ingredient of paper to this day. The word 'paper' is derived from papyrus, an early form of paper used by the Egyptians and made from a tall reed, *Cyperus papyrus*, growing mainly in the Nile Delta. Making papyrus was relatively simple, the stem of the reed being cut into strips, wetted, laid in two layers at right angles and pressed. In paper making, wood is pulped to release fibres which are put into a dilute solution that turns them into a slurry; this is then drained over a screen so that it forms a mat of randomly interwoven fibres. Not only does this make a stronger substance than papyrus, it allows for much more variation in weight and thickness. Papermaking today is a totally mechanised and immensely profitable business, though specialist hand-made paper is still produced on a small scale. Wood for pulp is in constant demand, and the cause of much deforestation.

Charcoal, once an essential fuel both domestically and industrially on a world scale though now much less widely used, is still a threat to woodland in Africa and South America. We may think of charcoal today as little more than a plaything, handy for firing barbecues or possibly sketching a nude, but it remains invaluable as an ingredient in gunpowder and a source of carbon in chemical reactions, and as a filter it lurks behind many of our modern gadgets such as air conditioners and kitchen extractors. This purifying property characterises the role of charcoal in the homeopathic remedy Carbo vegetabilis, made of beechwood. I also can't resist informing you that charcoal plays a vital role in 'flatulence odor control products'.

Trees are a primary constituent of peat and coal – whole swathes of forest in low-lying areas, with all the related vegetative matter, laid low and compressed over millions of years until mined, on this timescale very recently, by man. Wood and coal are intertwined in many ways: wood provided the shafts and props that made mining possible, and coal fired the Industrial Revolution – which in time made us just a little less dependent on wood. Which is just as well, since the wood is fast running out.

The trunk in its pure form as pole has played a part in the affairs of men, both practical and religious, since earliest times. I was surprised to learn that the totem pole of the North American Indians is more a teller of stories than an object of worship, its elaborately carved symbolic language recounting legends and disseminating news. Not religious either is the maypole, though entwined with many pagan rituals and festivities, many of Germanic origin but adopted, and adapted, throughout Europe. The British versions began in the fourteenth century and continue to this day, mostly in the form of rather chaste morris dancing on the village green on May Day. The trunk most commonly used as a maypole is that of the ash, a tree associated with spring and rebirth.

Flagpoles were traditionally made of single trunks, usually conifers, reaching heights of up to 70 metres – a figure more than doubled by modern freestanding flagpoles made of steel, but dwarfing the Trafalgar Square Christmas tree, a Norway spruce given annually by the people of Oslo in gratitude for Britain's support during the Second World War.

The Lombardy poplar* is an obvious choice as a pole, its trunk being straight, supple and naturally pole-like. In France poplar is grown in tightly packed, regimented plantations, each tree straining up towards the available light. It doesn't need spelling out why such trunks played a prominent part in fertility rites. In southern Spain a ceremony with Ibero-Celtic roots which continues to this day involves a votive offering of a poplar made to the Earth Goddess in return for the fertility of the young men of the village. On the Dia del Chopo, day of the poplar tree (23 June, eve of St John's or Midsummer Day), the young men set off in search of a worthy tree. When the largest and straightest has been selected, it is felled, stripped of its branches and carried back in triumph to the village – often up to three kilometres away – with much sweat and the need for frequent stops (and beer). The poplar, with the flags of Spain, Andalusia and the village tied to its top, is then erected and held in place against the church tower – which is ideally but not always dwarfed by the pole – in a symbolic if unintended union of the pagan and Christian.

**Lombardy poplars on Sheep's Green, Cambridge, an area of preserved meadowland close to the centre of the city.**

Tall and straight too is the larch, the chosen wood for the strange, and also sexually suggestive, Scottish sport of tossing the caber. This hefty trunk, usually about six metres long and weighing around 80 kilograms, is hard enough to pick up, let alone toss into the air as you and I would a matchstick. The caber is grasped in both hands, and the throw is judged not just on its length but on the way the caber falls. The judging is itself a skill, and the whole colourful exercise, complete with bagpipes and kilts, is a highlight of the Highland Games.

More serious uses for more serious trunks are as masts of ships. To look up through the rigging on an old sailing ship is like looking up through a forest, both awe-inspiring and dizzy-making. Although a single mast could be up to 40 metres long, as ship design improved even higher masts were needed, so a system of adding one to another was devised, with up to four separate pieces forming the mast. Sailing ships had at least three masts, some as many as seven, and tall, strong, straight tree trunks, most usually of conifer, were in great demand right up to the beginning of the twentieth century. But with the use of metal and plastic substitutes and ways of laminating and joining wood, these huge trunks were no longer needed, and those trees that escaped felling are now more likely to be venerated for their age and majesty, even protected. One species with a particularly dramatic trunk is the monkey puzzle, now almost exclusively grown as an ornamental. This exotic-looking conifer, with its large spiky leaves on drooping plumes that drop randomly and often spearlike into the ground, can grow to 40 metres with a trunk as straight as a die. There is one well over 20 metres high (we guess) in the garden where I live, its base grey and wrinkled like an elephant's foot, and on its otherwise immaculate trunk, one huge carbuncle like a weathered granite corbel that grins in the moonlight.

Trunks are distinctive, and most people will be able to recognise a tree from its trunk, if not the shape then the bark – its colour and texture. Though some barks may appear similar from a distance, all have subtle differences in texture and in the way they grow. The bark of the plane tree is unusually inflexible, which means that it is unable to grow with the tree, which in turn makes the bark peel off in large flakes. This has two advantages. The first is that it allows the plane to survive heavy pollution, which is why it is a favourite tree for growing in towns – there is even a hybrid called the London Plane. The other advantage is that the peeling bark reveals the most exquisite range of colours, a mottled mosaic in a subtle range of smoky blues, greens and ash-greys, each one unique, each a work of art.

Another tree whose bark peels, but in a different way, is the eucalyptus. This huge family of 700 species, loathed by many for their invasive, greedy habits, could take up the whole book (as they have taken up most of Australia), but even the bark comes in so many shapes and colours that it is hard to do it justice in a few words. Although there are types with hard bark, aptly called

**Rainbow eucalyptus is the common name of this eucalypt, but the Latin one is more explicit: *Eucalyptus deglupta*, meaning stripped or flayed.**

ironbarks, and others that are ribboned and tesselated, it is the stringy barks that excel themselves in their range of colours: the more sober ones of grey-blue and silver, others with flashes of mauve, turquoise and pistachio green, and some that look as if a painter has used them to try out all the colours on his palette.

Talking of colour brings us to that strangest of trees, the cork oak. Most trees will die if the bark is cut horizontally around their girth, since that cuts off the vital supply of water and nutrients carried up the trunk to fuel the leaves and flowers. But, probably to protect it from fire, the cork oak has evolved over thousands of years a distinctive corrugated bark with a thick spongy inner layer, which provides us with the cork that, amongst other things, stops our bottles and is so comfortable to walk on – though both of those uses are now diminishing as synthetics take over. The harvesting of cork, if a dying art, continues in Portugal and southern Spain, where the cut trunks, from warm ochre slowly deepening through every shade of orange, burgundy and umber until they become almost black again, are a characteristic of the landscape.

**A recently cut cork oak, Andalusia.**

The cork tree's unique ability to regrow its bark means that cork can be stripped from the whole trunk, from ground level up to and often including the base of the first main branches. Each tree is cut every nine or so years, and the cutting is still done by hand, a skilled job carried out within the course of a few weeks in midsummer by bands of itinerant cutters. The distinctive thwack as the razor-sharp axe hits the spongy bark, followed by something between a creak and a groan as the pieces of cork are eased away from the trunk – sounds as old as time – are the first one knows of the cutters' arrival. On the deeply valleyed hillside above my house in Andalusia, too steep and inaccessible for anything mechanised, the cork was brought down to the road on mules, in toppling stacks roped on to wooden panniers. Who knows how much longer this time- and energy-consuming practice will continue, and I feel privileged to have seen it.

It is to do with climate of course, but most of our native British trees have bark that is hard and frost-resistant, much of it ridged and nobbly, some deeply furrowed. But trees whose enemies are drought have different needs, and many trees are succulents. To Western eyes, some of these look more like giant garden plants, and the euphorbia is just that. Among the many members of this prolific family are spiny trees whose branches grow out of their trunk like a head of broccoli; these euphorbias flourish in Africa but are related to the leggy plants of our herbaceous borders, both having in common the milky fluid otherwise known as latex.

**Flowers of the Judas tree.**

Trunks produce many surprising things, from the clumps of beautiful flowers of the Judas tree to the shocking twisted spines of the honey locust tree that make barbed wire look cuddly by comparison. The euphorbias' latex is a hostile substance also, an irritant that can cause permanent blindness. But another type of latex, from the para rubber tree (also a member of the euphorbia family), provides us with innumerable useful things, from erasers to Wellington boots. Although natural rubber is slowly being replaced by a synthetic version, rubber production is still big business, mostly in South-West Asia but in South America and North Africa too.

**Defences of the honey locust tree.**

The rubber latex is in the bark of the tree, and 'tapping' it is a skilled operation since if the tapper goes too deep the tree's cambium layer, its food supply, will be damaged. A spout is knocked into the bark, and through this the latex drips into a cup – or a coconut shell – suspended on a wire around the tree. The latex runs for about four hours before coagulating, and the trees can be tapped every two or three days over many years.

A more mellow substance exuded naturally through the bark is resin. Different from sap, resin is viscous and strong-smelling, composed of volatile terpenes (from the word turpentine) which also provide the aroma in essential oils. Resin is sticky – so sticky that it gives its name to many types of adhesive, including the modern synthetic ones, all of which emerge from their containers in liquid form but almost immediately harden and become immovable. Natural resin hardens too, but lacking the catalyst of the artificial resins, more slowly – very slowly in the case of amber. But though talk of resin may bring back nostalgic memories of childhood model-making and a smell that catches you by the throat, it should also summon up the more pleasing aroma of incense. For resin is the source of frankincense and myrrh, those priceless gifts of the Three Wise Men, the main constituent of incense and used from much earlier times by the ancient Egyptians to mummify their pharaohs. The trees that bear frankincense and myrrh have their own family name, the Burseraceae, one of whose species, *Bursera simaruba*, has the delightful common name of gumbo-limbo and the more explicit nickname of the Tourist Tree, so called because its reddish bark peels like a sunburnt visitor to its native habitats in the Bahamas and South America.

Although as far as I know there is no bark cure for sunburn, the curative properties of bark are many. Most notable is probably cinchona bark, also called Peruvian or Jesuit's bark, from which we get quinine. This cure with its far-reaching consequences for treating malaria was discovered in Lima by a Jesuit apothecary, who brought the bark back to Spain in 1632, from where it quickly travelled to the rest of Europe.* Barks from other species of the cinchona, named unimaginatively yellow, grey and red bark, were also valued medicinally, principally for fevers but also for treating gangrene, scurvy, rheumatism and as a tonic. Of our own trees, the bark of alder was used to treat colds and chills, that of elm for broken bones and gout, and willow bark was chewed on to relieve aches and pains. And if all this sounds remote and too much like witchcraft, you may be surprised to know that bark still plays an important part in modern medicine. That willow bark – specifically the white willow, *Salix alba* – is the source of salicylic acid, from which aspirin is derived; and betulinic acid, obtained from birch bark and long used for its anti-inflammatory properties, has recently been discovered to have exciting potential as a selective cancer inhibitor. It is for more than their wood that tree trunks are valuable to man.

Bark is host to permanent residents, lodgers, visitors and various other hangers-on. Many of the insects whose livelihood centres on tree bark owe their common names to their activities, a busy horde that includes wood-boring weevils, carpenter worms, twig girdlers and, most damaging of all, bark beetles. Drawn by these come the birds, small ones such as crested tits, nuthatches, tree-creepers and others with engagingly acrobatic ways of reaching their prey, as well as the woodpecker family whose main concern is to get through the bark and into the trunk. These colourful birds, whose different species vary in size from 8cm up to 60cm, have evolved immensely sharp, powerful beaks that act as chisels, and the wherewithal to use these. The statistics are breathtaking: the beak hits the trunk at 20kph, and the slowdown at impact is a thousand times more powerful than the gravitational force and a hundred times higher than the acceleration of a spacecraft during lift-off, the head and neck being specially adapted to cope with this force and to prevent it from damaging the brain.

Less frenetic but almost equally colourful are the lichens and mosses that take up residence on tree bark. Lichen comes in a painterly array of shapes, forms and colours – wisps, layers, crusts and warts in silver, gold and russet. Particular lichens are attracted to certain trees, their presence reflecting different habitats and climates. Their morphology is complex and this is not the place to explore it, but no one who notices trees can be oblivious to the role of lichens in the appearance of trunk and branch; at times they can seem like pictures on an exhibition wall. Moss too, though this in its density is more of a blanket, draped over branches or hung against the trunk, a thick green comforter, sign of dampness and shade.

**It is easy to understand why the baobab is also known as the bottle tree.**

Trees in hot climates need water, and the baobab solves this problem in a unique way, by using its trunk as a storage tank – which is exactly what it looks like, the sparse branches sticking out of the top as if stuck into a vase. (It is also known as the upside down tree, as when bare – it is deciduous – its branches look like roots.) The extraordinary trunk of one of the Madagascar baobabs – where six of the eight species originate – grows up to 18 metres high and can hold 120,000 litres of water. But many of those familiar from moody sunset shots in Africa are squat and of huge girth, hunched like giant whiskery toads in the bare red earth. Because of its composition the trunk doesn't put on annual growth rings so cannot be dated by dendrochronology, but baobabs can live to a great age, often well over a thousand years. The trunk has other uses too: a white powder from the interior acts as a sweetener and thickener in traditional African cookery, and the hollowed out trunk as a meeting place for the village.

A tree with its focus at the other end of the trunk is the Moreton Bay fig, whose immense root system dominates both the tree and our impression of it, making it hard to know where trunk ends and roots begin. In fact there is no neat answer, for these are no normal roots taking their rightful place below ground, but swirling extrusions that appear to have flowed from the trunk they girdle in order to support these huge trees on the shallow soil in which they thrive. This tree, like others in the Ficus family, may also put down aerial roots, voracious feelers that drop from high branches so that the roots and trunk become indistinguishable in a gigantic tangled cavern. From here to the subject of roots is a small step.

**Although these are known as buttress roots they do not act as buttresses, whose force is external, but draw their tension from within the trunk.**

# ⥱ ROOT ⥲

Roots are the plant's worker ants. Their work goes on selflessly, mostly underground, grubbing around in the soil gathering nutrients and passing them on without looking for any recognition or reward. Tree roots have another purpose of course, that of literally grounding the tree, preventing it – not always successfully – from keeling over or taking off in a high wind.

This is true of the majority of trees familiar to us in the West, but there are trees whose roots are habitually flamboyant, such as the figs we have just left in the last chapter. The banyan, also a fig, has aerial roots that are parasitic, the seed germinating high in a host tree and sending down roots in search of the ground; some of these become so thick that they end up as independent trunks surrounding the host, which then dies; others seem almost to be walking on their roots, using them as stilts to spread out over large areas.

**Cork oak in the Sierra de Aracena.**

Although no competition for these exotica, trees with normally conventional habits will often manage to take root in whatever unlikely spot they happen upon. The enormous and very healthy cork oak pictured here seems to have poured its trunk over the boulder, with no apparent roots keeping it attached. But, completely exposed as it is, this fine old tree must have weathered many storms. In the grounds of my house in Brittany, more sheltered but equally impressive, a beech tree grows eight metres tall on a couple of roots spread over a granite boulder well over a metre high before finding their way to the soil below.

**Of common British trees, the beech is the most adept at spreading its roots above ground in search of anchorage and sustenance.**

Although most trees propagate through seed, some can produce roots from their small branches. Willows are the champions at this, sprouting vigorous roots wherever a branch or twig comes into contact with water. One can watch this happening by putting a willow twig in a glass container full of water; in a few weeks it will be surrounded by a mass of roots. And a bundle of cut willow left on the ground will produce roots from several places along each branch. It is this rooting ability that makes the willow one of the most used branches for 'living' sculptures and furniture. Another quick rooter is hazel, and often I have used a hazel stick, long and straight and ideal for the purpose, to act as a prop for a small tree or shrub only to find the hazel sprouting leaves and the supported one starved of nutrients and struggling.

The rapid growth and invasive habits of some tree roots may serve them well in the wild but can become a problem in our more confined spaces. Many a war between neighbours has arisen over that pretty little eucalyptus, planted in the back garden but now become a monster with roots invading fences, foundations and drains and with no respect for territory.

When a seed germinates, that miracle beyond all understanding, the first thing to emerge is a root: what goes on beneath the surface precedes any upward growth. In most trees the first root, the taproot, acts as a pointed bore, penetrating the earth and preparing the way for more tender ones that sprout from it. The carrot is one giant taproot that develops no more than the odd whisker (the radish, bulbous though it is, takes its name from *radix*, the Latin for root), but in trees the taproot is usually replaced by the smaller fibrous roots that spread out horizontally to form a dense network and eventually take over from the main root.

This upper layer of the root system is essential for spreading the weight of the tree and balancing its branches, and typically extends as wide as the tree is high. It is also surprisingly shallow and may not go much deeper than half a metre, with only a few roots going down into the subsoil and below to act as anchors. This enables the surface roots to get their sustenance from the rich topsoil, which they then pass on to the rest of the tree.

Which brings us to the subject of natural mulch, and the endless cycle by which the tree takes up goodness from the soil and then returns it, through the shedding of leaves and the eventual death and decomposition of the whole tree. It is all so effortlessly efficient, so economical, so logical – so ecological.

Not all trees behave in this way though. The habits of trees – of all plants – are dictated by both soil and climate, and the evergreens have a different calendar to that of their deciduous relations. But every tree, with the possible exception of the rogue eucalyptus, returns something to the soil, and it is in this rich mix of nutrients – remarkably similar to our own diet of proteins, carbohydrates, fats and sugars – that the roots thrive.

Because so many British trees are deciduous, our topsoil, the topmost layer of soil, seldom more than 20 centimetres deep, is in most areas rich in organic compounds and micro-organisms. The soil right at the surface, humus, is composed of organic matter as well as silt and sand or clay (sometimes both); it is a rich dark brown and looks almost edible. I recall swapping boasts with a friend about our respective topsoil: 'Mine looks like chocolate cake'. 'Well, mine is so sweet you could spread it on toast.' I also remember an Australian friend, down on her knees scraping up handfuls of humus, exclaiming, 'Oh, we just don't have any of this at home!'

For plant and animal matter to decompose into something useful, rather than just stagnate or rot, it needs oxygen, so this top layer of soil is correspondingly light. This allows the roots, which are delicate in their very early stages, to penetrate and push through the soil, to find their way around stones and into crevices, to explore and exploit the soil to their advantage. As the roots age they solidify while retaining their knotted forms, which makes them desirable both for their shape and their grain: sculptures often emerge from roots without any change needed to their natural form, and the intricately patterned roots of hazel and maple are sought out for marquetry and inlay.

**This natural sculpture may be a relic of the Caledonian forest.***

Many tree roots engage in some form of symbiotic relationship with fungi, usually that of mutual gain, though the word symbiosis, meaning 'living together', also applies to parasitism. Fine fungal threads, either enveloping or penetrating the roots, enable the tree to extract water and nutrients from the soil, in return for which the fungus gets sugar via the host's photosynthesis.

The result of a symbiotic relationship with tree roots, mainly those of the oak, beech, hazel and pine, are truffles. These subterranean mushrooms, witchy and warty and worth their weight in gold, come in several varieties, colours and flavours. Truffles have been treasured for many thousands of years and were mentioned in the twentieth century BC, though whether their value then equalled the current world record price of US $330,000 for a 1.360 kilo (3lb) white Italian truffle we will never know.

Below the rich goodies of the topsoil lies the subsoil, more dense and substantial, made up of the same minerals but lacking the organic matter. The main tree roots, which need to go a lot deeper than the topsoil, will get much of their supply of water and minerals from the half metre or so of subsoil, as well as being able to keep the tree stable on this more solid base.

It is clear that the soil will be affected by both geology and climate, but this is in danger of getting too technical. So instead of going deeper into the soil, let's take a more poetic look at roots.

Root metaphors abound in our language: ideas take root, we endeavour to get to the root of the matter, prejudices may be deep rooted and, as the Bible reminds us, 'Love of money is the root of all evil' (not money itself, note, but the love of it). I have always found the expression 'root and branch' baffling, since it omits the all-important part holding the two together. To root out the offending object – literally to eradicate it – seems an altogether more satisfying thing to do, and visual too as the mattock thumps and the sweat flows. For without its roots the tree will die.

Because of their largely subterranean lifestyle, roots don't feature too much in art. Magritte's *Le Domaine Enchanté* (illustration page 68) disturbs and provokes in the way of all his images, its seeming

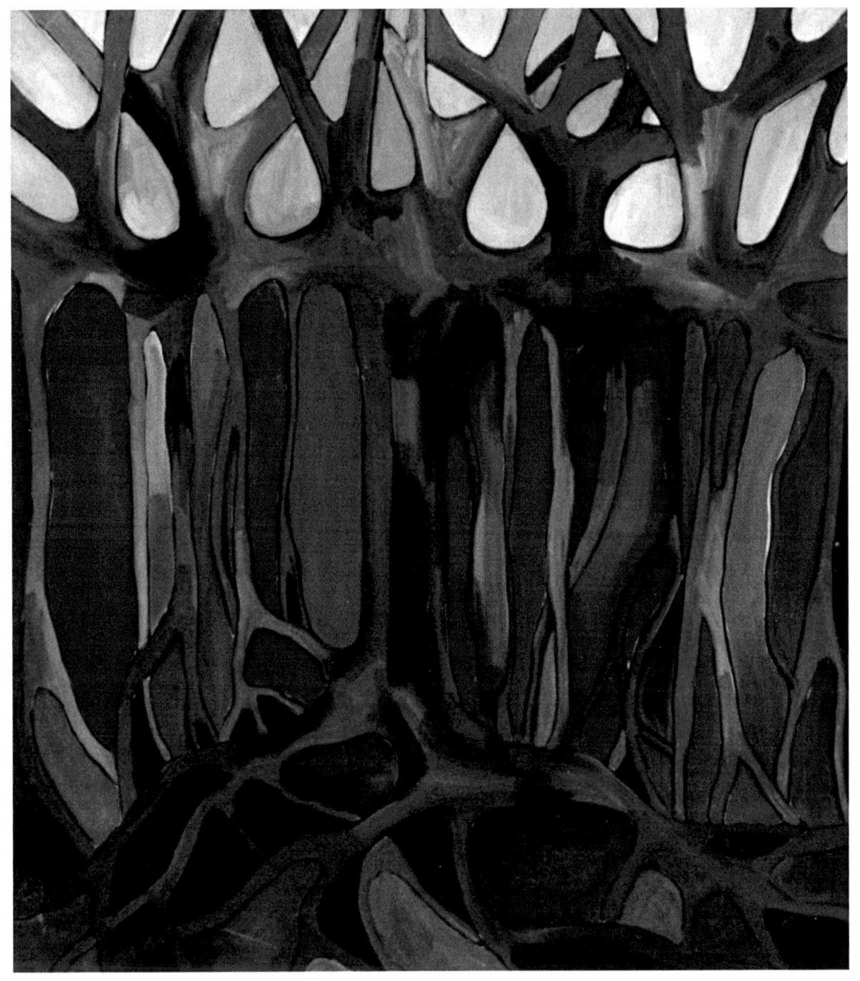

**Bharati Chaudhuri, *The Tree*, 1973.**

simplicity leaving us wondering, confused. What's going on here? What lurks beneath the obvious? A contemporary Indian artist, Bharati Chaudhuri, confuses us in a different way. In her painting inspired by the strangler roots of the banyan, she has created a 'middle earth' between soil and sky which is straddled by the roots. In some depictions of the Tree of Life, especially the Norse Yggdrasil, the roots are the mirror image of the branches, the whole enclosed in a circle. And in others this circle represents the world, with Earth perched on a tree in the middle like a cake on a stand.

**Flora or fauna? Beech roots at Glastonbury, Somerset.**

But most artists focus on those roots that lie on the surface, the tangled knotty ones that look like escapees from a zoo. This beauty in ugliness is compelling. Looking at the roots pictured here we ourselves writhe; we creep with them across the surface, searching for some way of getting down into that nourishing humus of the Underworld.

Most if not all of the world's religions and mythologies feature a place under the ground where stories are told and lives imagined. From Cerberus, the many-headed dog of the Romans and Greeks who guarded the entrance to the Underworld, and Hercules (or Heracles) sent on his final labour to capture him; or Pluto, otherwise known as Hades, ruler of the Underworld, and Proserpine (Persephone) its queen – these figures from classical mythology thrive beneath the ground. In the myths of people as diverse as the Phoenicians and Slavs, Siberians and Aborigines, the Underworld plays a significant role, though it took Christianity to associate this with the Inferno and eternal damnation. And Alice's Wonderland is of course beneath the ground. It seems we humans too have some business down there.

Writers have taken roots as symbolic of connection with these other forces, as well as metaphors for our background, where we belong. Alex Haley's book *Roots* and its television adaptation were hugely successful as well as controversial; the story, part fact and part fiction, traces Haley's own family back to its protagonist, his African forebear born in a village in Gambia and sold into slavery in 1767.*

In a haunting image, the American poet Marguerite Wilkinson writes:

> *A woman's beloved is to her as the roots of the willow.*
> *Long, strong, white roots, bedded lovingly in the dark.*

But Dylan Thomas must have the last word, in a much used quote that usually peters out half way through:

> *The force that through the green fuse drives the flower*
> *Drives my green age; that blasts the roots of trees*
> *Is my destroyer.*

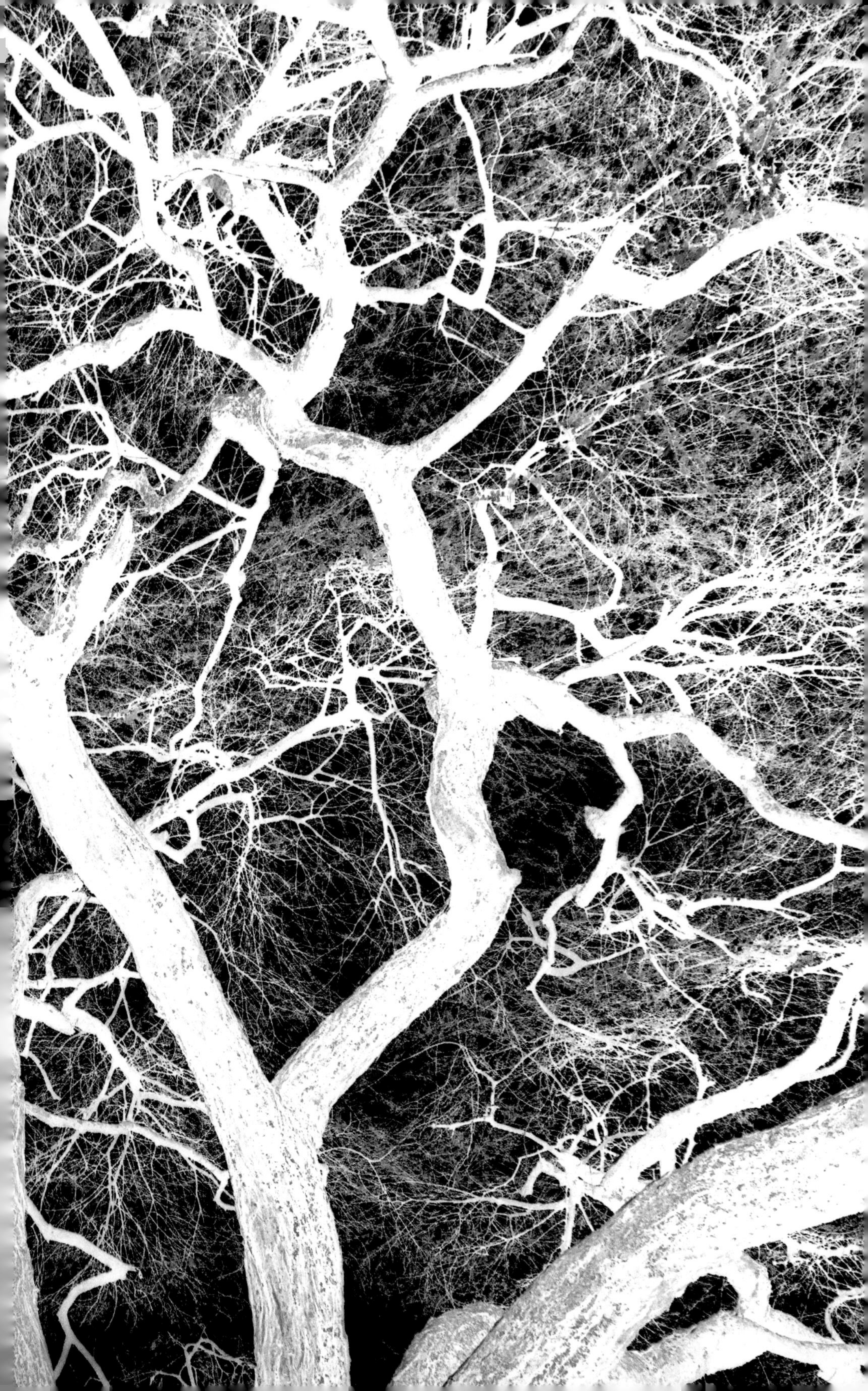

# BRANCH

The force that drives the flowers of a tree flows through its branches. These arms on the torso of the trunk – often called limbs – which shimmer off into a myriad tiny fingers, are as essential to the 'end product', the leaves, flowers, fruit and seeds, as is the trunk; they are merely extensions of the trunk. And although there are other names for them according to thickness – bough, twig, twiglet – it is the branch that gives its name to any number of metaphorical uses, whether railway line, river or (money) bank.

If a tree is recognisable from the trunk, it is equally so from its outline: the shape, form, density and deployment of branches make up its character, its aura. Given the basic facts of trunk and branches – forgetting about the leaves for a moment – it seems astonishing that the shape of a tree can vary so greatly, be so distinctive. Yet few people who know anything at all about trees, or keep their eyes open to natural forms, would mistake a beech for an elm, or a lime for an ash. Easier not to mistake a yew for any of these, but how about a young yew and a small squat fir tree? Yet somehow they are different.

The overall form may be strictly upright, as in the Lombardy poplar; upright and rounded, as in the beech; rounded but more mushroom-like, as in the oak; a figure-of-eight like our sadly missed English elms; and all of these shapes are defined by their branches, clearly noticeable in the bareness of winter and only slightly blurred by the addition of leaves. The same goes for the evergreens, though their form is more camouflaged, and influenced, by their dense, ever-present foliage.

A branch can be anything from a bough as thick as another tree's trunk to twiglets as fine as lace. Its bark will have the same basic characteristics as the bark of its parent trunk, though because it is smaller the texture and sometimes the markings may differ. And the individual form of the branches, both large and small, differs from tree to tree just as that affects the overall form of the tree.

**Bare branches provide a perfect example of 'inscape', unique form.***

It is not only the characteristics of branches that differ, for they can be rounded, pointed or spiked, straight or knobbly, tightly crabbed or flowing, orderly or chaotic; they are also defined by their way of growing – upward reaching, outward going, clumped together or spread out, rigid or waving or weeping. Over and above the differences, however, are some common traits, the principal of which is respect for each other's space. Or possibly it is less public-spirited than this, more a selfishness that motivates each small branch to find the best position in relation to its neighbours for getting the sunlight and moisture and pollinators that its leaves and flowers need.

To trace the slow growth of a protective scab over a well-cut branch is to marvel at nature's self-healing ability. Fleshlike, the cambium layer surrounds the cut, swells and then begins to close round it so that over the years the tree's open wounds become its noble scars. This only happens if the cut is clean, which is why pruning is a skill, one that can never be matched by a machine, for the needs, the necessary angle of each cut are individual. That professional tree cutters are called tree surgeons is no accident, and because they are cutting 'limbs' we identify with the process. Which is why mechanical cutters cause so much damage to the trees, and suffering to any observer who cares about them. Sentient or not themselves, surely the sight of mangled trees in a hedgerow, their limbs indiscriminately torn, wrenched and minced, must make one want to cry out. More recent cutters have improved, in that the cut is cleaner and less destructive of the tissue, but still the angles are usually wrong so that the tree ends up bristling with stumps, the beauty of its natural form irretrievably lost.

Here in Brittany trees, mostly beech and oak, are planted on high banks to form giant hedges that act as both windbreaks and boundaries. In England mixed hedges are an integral part of the landscape; although primarily formed by shrubby species such as hawthorn and hazel, adorned by wild rose and thickened with bramble and ivy, they often include some large trees to give variety and added shade. Laying hedges by hand is a skill, now a dying one mostly confined to local enthusiasts who run courses and arrange competitions. Having grown up with laid hedges in Oxfordshire I was surprised to learn how many variations of form and method there were around Britain. Briefly, hedgerows are limited in size by having the trunks cut diagonally (this is the skilled part, as the cut must be deep enough to allow the laying yet not so deep as to cut off the food supply) and then laid at a lesser diagonal along the line of the hedge. The branches are then woven into this to form a dense living barrier against livestock, predators and the elements. Some thinning and cleaning of branches was often called for in the interests of the structure of the hedge and depending on the animals being contained, and different styles in various parts of the country involved the addition of support stakes, singly or in pairs, and binding in different patterns with flexible branches such as willow or hazel. That these hedges can withstand the onslaught of a field of frisky young bullocks is testament to their strength, and their beauty is beyond doubt.

**This professionally laid hawthorn hedge in Beaconsfield has ash standards and sweet chestnut stakes, with hazel binders in a style common to the Midlands.**

An ancient and enduring way of farming wood is coppicing. In this the branches of certain trees, most often willow, hazel, sweet chestnut and ash, are cut on a regular basis to provide poles of a size suitable for fence posts, charcoal, furniture or firewood or, if cut younger as in the case of willows, for fence-making and weaving. The branches are cut at ground level, leaving behind the parent stump, called a stool, from which sprout more young shoots in a cycle of several years depending on the tree and the intended uses of the branches, in a process that continually rejuvenates the stump and can continue for hundreds of years.

Small branches become sticks and larger ones logs when they are dead, whether through natural causes or man's intervention, and the main use of sticks and logs is for burning. Before the days of firelighters, everyone – particularly the young and old, as this was considered an easy job – would go out foraging for kindling, usually brought back in bundles on their backs, sometimes on the back of a donkey or mule, to be stacked and kept dry for the important job of lighting the daily fire. 'He who would live longest should fetch his wood furthest' went a moralistic though probably well-founded saying.

Larger branches used for firewood, either cut specifically for this or salvaged from fallen trees, need to be dried before they can be burnt, for at least a year and preferably longer. Wood varies considerably in its heat production and efficiency, from oak which burns long and hot to elm which burns 'like churchyard mould' and poplar which can hardly be persuaded to burn at all. Wood fires have provided our ancestors with heat, light and cooked food for as long as time, since the discovery of fire by Early Stone Age man some 800,000 years ago – which of course came about by rubbing sticks together. The fascination for fire is thus deep in our genes, even if we live in a centrally heated house in a city and seldom come across raw fire, as was evident from a primetime television programme put out in Norway in 2013 which ran for twelve hours and featured a burning log fire with very little commentary. This atavistic compulsion to watch the moving flame – now most often replaced by the flickering television screen – is what draws us to bonfires, the last remaining

**The power and unpredictability of fire are part of its fascination.**

primitive celebration of fire in our tame environment. (The word bonfire comes from 'banefire', the medieval burning of bones.) Not many people can resist the lure of a bonfire, and one made of the prunings of large trees can be exciting, even potentially dangerous as the wind suddenly changes direction, or the fire takes hold and creeps towards other trees or the garden shed.

Wood fires bring out the poet in one, as described by Adam Foulds in an imagined scene in the life of the nature poet John Clare: '. . . he looked and realised that those were particular logs being consumed, logs from particular trees burning with particular flames in that exact place at that specific hour and it would only ever occur once in the history of the world and that was now. Birds had landed on them, particular birds, and creatures had crawled across them, light had revolved around them, winds swayed them, unique clouds passed over them, and they would be ashes in the morning.'

I once saw the bonfire to end all bonfires, made not only of mere branches but of whole tree trunks, massive old beeches felled within the Iron Age fort of Danebury, in Hampshire. Because of the structure of the ring with its ramparts and ditches the trees could not be removed, so the only option was to burn them in situ. This was before the time of clever lifting gear, or of warning signs and guided tours, and I and my children stayed up on the fort until after dark, watching this bonfire of the giants as they faded and slipped into their own rose-grey ashes.

**Old pollarded willows in their favourite site, close to water.**

Sticks have other uses too. As sticks and staves, to beat recalcitrant donkeys – and children, though for them the more usual weapon of choice was a cane, made of whippy willow. But more benignly the stick may become a baton, wielded by a conductor in full evening dress, or passed on in a relay race; or the wand of the Fairy Godmother, instrument of wishes come true. And as a rod, the willow or the hazel can tell you where water lies under the earth's surface, its twist in your hands as unmistakable and uncontrollable as if a small animal were cupped there.

Hazel sticks were used as thatching spars, barrel hoops and as the wattle of wattle and daub – the daub being mud and lime – in the old humble method of plastering. But as well as fulfilling these utilitarian roles, the beautiful bronze-patterned hazel was highly regarded as having magical properties and was used as a staff by the Druids and by pilgrims, and continues to be the favourite wood for walking sticks to this day.

The Celts collected yew sticks, casting them on the ground and reading the patterns they made as divination. Yew was also the favoured wood for forming the sacred glyphs of the ogham alphabet, also known as the Celtic Tree Alphabet, twenty letters formed by straight lines, each based on the Irish-Celtic name of a tree and each having an equivalent runic symbol. This simple but effective writing system was used to convey messages as well as for divination, and was later inscribed on wood or stones.

But back to more mundane uses, and to the willow again, invaluable in its straightness and flexibility for weaving – not only for the baskets we associate with that, but for making coracles, fish traps, windbreaks, hurdles and a host of other useful things. For this the willow is cut very young, when the golden withies are still supple, and to ensure a supply of these the trees are pollarded – a process similar to coppicing except that the tree trunk is allowed to develop and the young branches are harvested from this rather than a stump at ground level. Willow is still grown this way, the withies being cut every few years, but pollarding is now most often seen in towns and parks where trees are controlled and kept small by this harsh and at times unattractive method of pruning.

Bundles of sticks were used in various ways, as faggots for firewood, as brushwood harrows (hawthorn), as brushes and brooms, the names so familiar that we probably don't associate them with their origins. From faggots comes the more sinister word 'fascism', its symbol a bundle of rods tied round an axehead, adopted by the Romans as a sign of authority and later as the emblem and name of the totalitarian movement in Italy during the First World War. The symbolism of the faggot is 'strength through unity', a concept supported by anyone who has ever tried to break a bundle of sticks.

Sticks of differing sizes were used to make charcoal, as well as the larger wood needed to fire furnaces and forges. Charcoal sticks – mostly of willow – and powder have long been in use as artist's materials. The early Renaissance fresco painters, having made their preliminary sketches in charcoal, then perforated the design and transferred it on to the wet plaster by means of 'pouncing', dabbing with a small porous bag filled with charcoal dust. In a technique called parsemage, invented by the twentieth-century surrealist painter Ithell Colquhoun, charcoal dust (or coloured chalk) is scattered over the surface of water in a small tank and skimmed on to stiff paper or card to create random images. David Nash, whose sculptures of charred wood evolve in the making, reverses the usual process and makes drawings of them in charcoal after they are completed.

Serious artists as well as amateurs make sculptures of sticks and driftwood. Henry Moore collected driftwood and odd pieces of wood that caught his eye, and used them as inspiration for much larger works later cast in bronze. Andy Goldsworthy makes sculptures of both living and dead wood, some on a huge scale like those in Grizedale Forest. Heather Jansch uses driftwood to make astonishingly lifelike and lively horses that prance on the seashore in a thought-provoking cycle of land-water-land-water; some of her pieces too are now cast in bronze.

So far we have neglected the evergreens, whose year-long mantle of greenery, much of it dense and dark, makes the branches less noticeable but whose shape is nevertheless dictated by them.

The conifers, a huge group of cone-bearers which includes cedars, cypresses, firs, pines, spruces and yew, tend to have upward- or outward-facing branches bearing their cones aloft: as a generalisation, they are likely to be cone-shaped themselves. This

**The cedar of Lebanon, here seen flourishing in its native habitat.**

shape means – or demands – that their branches are supple, since they are defying gravity by waving their arms up at the sky. It also makes them vulnerable to heavy snow or very high winds unless they are tightly packed to support each other, particularly as they tend to have shallow roots. But the advantage of this exposure comes at pollination time, since almost all conifers are wind pollinated.

Never having lived in predominantly conifer country I don't have an eye for their form. Although I know from Van Gogh's paintings that some cypresses are particularly dark and pointed, and that the rapid growth of the wicked Leylandii has caused many normally tolerant people to fall out, I couldn't easily identify a cypress or distinguish it from a cedar at a hundred paces. Familiarity with trees is essential, but it's never too late to learn.

The cedar of Lebanon, a familiar sight on the manicured lawns of grand houses in Britain with its flat, widely spaced branches, only gains this characteristic horizontal sweep after many years, I discover, starting life in an upright spearlike form. Most of the firs are upright, sometimes pointed, but more spiky in outline than the cypresses; their branches have to be strong to support the heavy needles and large, erect cones. Spruces are always pointed, and their needles very short and sharp, with cones that hang downwards. Many species of pine are pointed too (though some are umbrella-shaped) and their needles are characteristically long, growing in what look like bunches but are actually spirals – in fact branches, needles and the scales on the cones are all arranged spirally. Yews are more easily recognisable, being densely black with a jagged outline and a hefty trunk. And in autumn our only deciduous species of conifer, the larch, is instantly recognisable through its downward-sloping branches and startling yellow-gold leaves. But as the eight families that come under the heading of conifers have hundreds of 'branches' – 630, to be precise – each with its own very subtle distinctions, identifying them is a challenge to the amateur. So let's just celebrate their differences and enjoy them for their form, their colour and the contrast these provide to our rounded deciduous trees.

From bough to branch to twig, the tree's limbs fan out and are reduced in size as they increase in number, all energy straining towards the tips. As already noted, the way in which they do this is unique to each family, part of their signature. Some grow gracefully, others less so, but all perform their function of providing the best possible circumstances for their offspring, the leaves, flowers and fruit. This is much easier to see in winter, perhaps against a pearly evening sky with every tiny twig etched in its finest tracery.

**Form and perspective, colour and texture.**

Twigs are distinctive in their structure too, and in the way they produce their buds, the way their leaves are attached and hang. But even in winter, dormant and apparently lifeless, they are

**Aaron Johnson, *Cypress Cove, Point Lobos*, two-colour woodcut, 2006.**
**There is something particularly pleasing about a woodcut depicting its source – 'a give and a take' as the artist puts it.**

distinguishable by their form, so that picking up a fallen twig one can tell immediately if it is beech or ash, lime or elm. Some are thick and chunky, others fine and tapered. Much of this is to do with function. The fat buds and heavy, bunched leaves of the horse chestnut demand sturdy twigs to support them, whereas the dainty buds and light leaves of the beech make no demands on their delicate twigs. Ash twigs are distinctive in their straightness, with tight, pointed almost black buds set alternately, while the sycamore's straight twigs have green buds opposite each other. Elm twigs zigzag, birches bend elegantly and field maple gawkily – all are different, every one is unique.

# LEAF

'The treees are coming into leaf like something almost being said', wrote Philip Larkin. New leaves: the aura of a tree shifting as its outline softens to a haze, and then the budding, the unfolding, the astonising greenness as the leaves emerge, slowly at first and then bursting with exuberance – waxy, innocent, as yet unblemished. This yearly miracle of spring calls forth something in us; we too feel the stirring, the disturbance of growth, the possibilities ahead. As we grow older it makes us take stock, seeing our own life-cycle unfold, wondering what the future holds, knowing that it is no longer limitless.

Leaves come in many shapes and sizes, but the one thing most have in common is their greenness. Depending on the pigments present it may be a yellow-green, as in the ginkgo, or have pink or orange tinges like the maples, or be as densely, darkly green as the yew, but the greenness, the chlorophyll, is there. This brings us to the subject of photosynthesis, a word familiar to every schoolchild even if he or she might struggle to explain what it is. Put very simply, photosynthesis is the leaf's way of turning light into energy into glucose, and the vital facilitator in this is chlorophyll, literally 'green leaf', a pigment that allows the plant to absorb light in the first stage of this process. (In trees such as the copper beech, which have significant amounts of other pigments, the chlorophyll may be masked, but the energy these pigments draw from sunlight is passed on to the chlorophyll for processing in the same way. The dark pigment is more prevalent in the parts of the tree exposed to light; in more shady areas, the leaves will be greener.)

The size, colour and form of leaves is dependent on both their habitat and their habits, the latter governed by the former. The thin tough spikes of the conifers, the fat tough paddles of the laurels, the sharp blades of the evergreen oaks, the shimmering leaves of the birches and poplars – all of these, and many more, are in direct response to

**Copper beech leaves showing variations in colour according to the light.**

their environment and their consequent needs: for light, for water, for sun and shade and protection from frost.

The way in which the leaves grow, their pairings and opposites, whether they stand stiffly to attention or droop seductively, whether they whisper, rustle or scrape, is also contingent upon their form. Some trees, like some people, are effortlessly graceful – think of the silver birch, the weeping willow. Robert Frost, who wrote simple poems about the trees in his part of New England, gives this haunting simile of birches 'trailing their leaves on the ground / Like girls on hands and knees that throw their hair / Before them over their heads to dry in the sun.' As with all of nature's designs, the leaves 'fit' their parent trees: the shapes and colours and forms of trunk, branch and leaf are complementary and harmonious. We would not have them any other way.

Leaves have many uses beyond the immediate interests of the tree: as fodder for insects, animals and occasionally humans, as shade for them and for other plants, as medicine, balm and dye. But their usefulness does not end there. Leaf mould is one of the main ingredients of the topsoil layer that all plants rely on for their nutrients, and the falling of leaves is a necessary part of the yearly cycle of decay and growth. Deciduous leaves, most of which fall as soon as the leaves have dried (beech and sometimes oak being exceptions), break down quickly, and given frost and rain will have become an indistinguishable brown mass by the following spring, slowly to be absorbed into the topsoil, helped by the insects, birds and animals that work it over in their different ways. The spikes and paddles of the conifers and evergreens have a different story. Their falling is continuous throughout the year, and the leaves are so well protected that breakdown is a very slow process, more of a grinding than a rotting. This quality of course enables evergreens to survive in harsh conditions at either end of the temperature scale, but it does not provide as rich, and certainly not as quick, a humus as the deciduous leaves. Somewhere in between, though 'evergreen' in name and habit, are leaves such as eucalyptus and holm oak, conventionally leaf-shaped – that is, elliptical – but becoming very hard and brittle as they dry, and even more obdurate in their dissolution.

Without the leaves of the white mulberry tree we would have no silk, for they are the sole food of the silkworm – nor indeed would there have been a Silk Route, that much travelled road from East to West that brought with it so many other materials and cultures. Silk production, and consequent mulberry cultivation, goes back four thousand years, to China, where silk acquired such value that it became a form of currency. Despite modern man-made textiles silk is still highly prized, and the amount produced continues to rise – indeed doubled in the last thirty years.

On a more mundane level, mulberry leaves feed several species of moth, some of the many hundreds of insects that rely on tree leaves for their sustenance. As gardeners, or especially market gardeners, we tend to regard leaf-eaters as pests, to be eradicated by fair means or foul. But they too have their weapons, developing methods of camouflage such as turning themselves into moving leaves, like the leaf insect, or the giant swallowtail larva, ravager of citrus fruit leaves, that resembles bird droppings in order to fool predators. Others just win through sheer numbers, and are capable of completey defoliating

**The camouflage of leaf insects is exact, down to their ribs and veins; some have even adopted a rocking walk to mimic a leaf in the wind.**

vast areas of forest; it was estimated that in 1981 over five million hectares of hardwood trees were stripped in the eastern United States alone. The leafcutter ants of the Amazon do not eat their spoil in situ but chomp the leaves off and drag them to their underground chambers where they are crushed into pulp in order to grow the fungus on which the ants feed. And most summer visitors to the Mediterranean will have noticed the white candy floss attached to pine trees without realising its nasty implications; for the caterpillars that emerge from these cocoons, climb down the tree and process nose-to-tail in a line like a piece of moving string have toxic hairs that cause acute distress to any animal or human investigating them. My dog nearly died from doing so; she frothed at the mouth, developed a high fever, and her tongue, which turned black, never recovered its shape after being eaten away at the edges. The trees too suffer, and can die from a severe infestation of these sticky webs.

But not all insects are destructive of their host. Trees have many ways of attracting insects, through their buds, their flowers and their leaves, that in turn scavenge pests such as aphids and scale insects. And as the bugs come, so do the birds. One of the many joys of being surrounded by trees is the accompanying bird life, and often a tree will seem to be alive with birds as a flock finds a particularly luscious supply of insect food. Sometimes they just sit on a branch and sing their hearts out.

Grazing animals – and not only goats – enjoy eating leaves, especially at times when the grass is not abundant. Deer, cattle and horses reach up and trim the leaves and small twigs, leaving trees with the straight manicured line often seen in parkland, at the height of whatever animals are grazing there. In the old days when there was a shortage of fodder, farmers would gather leaves for their stock; elm was particularly desirable, to the extent that four thousand years ago overharvesting of elm leaves led to a serious decline in the elm population. Holly leaves too were given to cattle, and believed to improve the taste of the milk. In southern Spain it was chestnut leaves that kept my donkey going at the end of a long harsh summer.

I was surprised to find how many leaves are edible for humans, even if some of them sound like hard work. Lime leaves seem to be the favourite; as well as making a soothing tea they can be eaten raw in large quantities, and have a mild taste. The hawthorn was known as

**Leaves provide roughage as well as nourishment for grazing animals.**

the 'bread and butter tree', its leaves said to be as nourishing – though surely not as tasty – as just that. Elder leaves too are recommended if eaten young, and if you don't like the taste you can scrunch them up and use them as an insect repellant.

Leaves play a large part in natural medicine, the most fashionable now perhaps being from that strange tree the ginkgo. Known as the maidenhair tree, after the fern of that name, it has leaves that would not look out of place on the sea floor but the tree, we are told, is a direct throwback to fossils of 270 million years old. These leaves, which contain flavonoids and terpenoids, are now processed into tablet form and sold by the million, believed to enhance concentration and memory and help in the treatment and prevention of dementia.

From Australia comes tea tree oil, a strong-smelling camphorous oil made from the crushed leaves of the melaleuca tree, inhaled for coughs and colds and used as an infusion or poultice to treat wounds and skin ailments. Another tree with a big reputation medicinally is the neem, known in India as 'the village pharmacy'. The leaves are recommended for ailments as diverse as anorexia, nosebleed, ulcers and eye problems, but there is hardly a part of the tree that is not cited as curing something, even if only intestinal worms. *

Pine needles have several medicinal applications, principally as a decoction for respiratory problems – which is perhaps also why pillows stuffed with pine needles were recommended as a cure for breathlessness, though they must have been extremely prickly to lie on. Rheumatism and skin diseases were also treated with pine, and pine needle tea is rich in vitamins A and C and was therefore taken to prevent scurvy when oranges were not available.

Dyes are made from several parts of the tree too, from heartwood and bark to berries, flowers and leaves. It's not surprising that the brilliant red of sumac leaves yields a dye, as well as being the source of the tannin traditionally used to treat Morocco leather, a particularly soft leather once prized for bookbinding, gloves and shoes. Henna, a hair and skin dye, comes from the ground leaves of a small tree of the same name but can also be made from walnut leaves. Henna's reddish-orange pigment gives a lift to dark hair, turning it chestnut – a name also applied to a horse's coat, rather paler – from the glowing colour of the newly opened nut.

From the alder come three different shades of dye: light brown from the leaves, green from the catkins and red from the bark, representing the elements of earth, water and fire.

For deciduous trees, spring and autumn are the seasons of high drama, with summer a welcome break where the fruits of all that activity are enjoyed, and winter a period of total dormancy.

**Chestnut leaves emerge covered in silvery down.**

New spring leaves come out as if from the caul, as wet and waxy as a newborn lamb. Their colour is so fresh you can taste it, like mint. Some are bronze or pink in their new state, before growing into their more sober adult green. The buds themselves are often bronze too, and sticky with resin to catch possible marauding insects. Chestnut trees are known for their sticky buds: bulbous, shiny, edible-looking lollipops that seem to be about to burst into leaf before one's eyes.

**This poplar bud appears to be giving birth to the crinkled leaves that will soon become sleek and elegant.**

Also sticky are the poplar's buds, and many have strong healing properties. As well as containing salicylates, the main ingredient of aspirin found in and named after the willow (see page 37), the buds – both leaf and flower – of the western balsam poplar or black cottonwood are the source of a healing ointment, the Balm of Gilead, and of various tinctures, salves, and oils rich in antioxidants. ('Balsam' is a rather vague word used to describe the smell of the resin of various plants; the distinctive sweet smell wafts from balsam poplars in springtime.)

The new leaves of the sycamore, like its related species the maple, may be dark and tinged with red, or yellow blushed with pink-gold; either way they are as innocent as all freshly hatched leaves, their flower buds a bright pink, the flowers dangling so daintily. But the seedlings, spread by their helicopter seed pods, are so prolific that one must harden one's heart and yank them out or they will quickly take over any available space.

Poplars, which turn bright yellow in autumn, begin life with a rosy hue. Springtime on the Seine is synonymous with the pink flush of newly emerging poplar leaves, and the French Impressionists liked to show off their ability to capture the shimmering of these leaves. Monet painted a series of poplars on the river Epte near his home in Giverny during the summer and autumn of 1891, and they were a favourite subject too for Sisley. Van Gogh, a few years earlier (1885), had painted *Lane with Poplars*, a sombre autumnal scene with poplars whose rustling leaves you can almost hear.

Perhaps the most delicate and seductive of all spring leaves however are those of the beech: the palest green, edged with down, transparent in sunlight, dancing on their widely fanned branches to catch all the available light, they epitomise new life and innocence. Springtime beeches en masse, as in the New Forest or the Pyrenees, are a sight never to grow tired of and never to be forgotten.

Mature leaves are, in the way of maturity, more stable; they have got over the hysteria and excitement of youth and settled into the serious business of representing their species and fulfilling the tasks expected of them. The violent, exploratory colours are now more subdued, and with a few exceptions like the poplars, the early flutterings have become more solidly based, more *dependable*.

The shapes of mature tree leaves are in essence surprisingly few, though their arrangement provides a much wider variety, and their substance yet more. There is a distinct difference between deciduous and evergreen leaves, the gap being bridged by the evergreen oaks and the eucalypts which are harder and more brittle than their deciduous relations but not typical of evergreens either. What is striking about the leaf shapes is how like trees they are.

**René Magritte, *Le Domaine Enchanté*, 1957.**

Keen botanists have subdivided these basic shapes and come up with no fewer than 63 headings, each with their Latin name. But without getting into any of that we can illustrate the variations on a theme by looking at the simple form of the elliptic leaf, the basic boat-shape that most children will go for when asked to draw a leaf. On the sweet chestnut, for example, these are quite deeply toothed and grow alternately on the stem, any number at a time. Ash leaves of similar shape are only slightly kinked at the edges, and grow opposite each other, usually eleven to a stalk, never more. The rowan, or mountain ash, are toothed but more rounded, growing opposite each other in varied odd numbers, always with one at the tip. Most willow leaves are an elongated version of this oval with a finer point, but the bay-leaved willow is back to the basic boat shape, growing alternately on a long stem. The horse chestnut, at first glance so different, is composed of seven slightly fatter ovals, varying hugely in size and all joined into the stem at the base, like a fan. All of these we are considering in outline, but each has subtle differences in texture, top and under surface, patterning and prominence of veining, and colour, which give them their particular character.

'The annual death of chlorophyll is a wonder to poetry and science alike: millions of tons of green pigment destroyed in a blaze of colour that sweeps down from the Arctic tundra, through the woods of north America and western Europe, at about forty miles a day. From this ritual dissolution, called senescence, the plants salvage sugars and proteins and stow them away in seeds and tubers. In the leaves, the orderly dismounting of molecules breaks down the chlorophyll to simpler parts, leaving yellow carotenoids and red or purple phenol to a brief, biochemical glory.' So writes Michael Viney, whose descriptions of the landscape on the northwest coast of Ireland combine gentle science lessons with a fine poetic imagination.

Senescence (from the Latin for old, *senex*, and from there via Jung to the archetype of the wise old man) is not just degenerative but a highly ordered process regulated by special genes for the transportation of the nutrients accumulated during youth and maturity – the psychlogical parallels being obvious. Autumn is thus as active a period as spring, a harvesting and preparation for winter.

**The many varieties of Japanese maple have been bred for their dramatic colours. This full-moon maple was photographed at the National Arboretum, Westonbirt.**

And those colours. There are the really flamboyant ones, of course, the maples and sumacs of crackling bonfire reds setting whole hillsides on fire, but to my mind a mixed woodland that goes through all the shades from palest yellow to darkest red is the most satisfying – as are the names, earthy and metallic: siennas and oxides, cadmiums and ochres, and so many shades of yellow with names like barium, uranium, strontium and zinc. (As I write, my eye is caught by a robin sitting on a bare branch on the wintry tree outside my window, its russet breast an afterthought of autumn.)

Larches, those odd deciduous conifers, proclaim their difference unmistakably: their drooping branches defined by mustard-yellow needles flare amongst the dark firs and pines, and soon each will stand in a golden pool as its needles fall around it.

Some leaves lose their chlorophyll evenly, but in others it leaches away from the tips leaving a pattern of different colours: each leaf its own individual autumn palette. And others will lose their flesh but retain the skeletal form of their veins, exposed in their lacy intricacy.

A few deciduous trees, beeches in particular, hang on to their dead leaves throughout winter, especially in sheltered places which is why they are also useful as hedging; but for most the brief spell of glory ends in the fall that gives its name to the season. The exposed beech loses its leaves like any other:

*Nakedly muscular, the beech no longer regrets*
*Its lost canopy, nor is shamed*
*By its disorder. The tatters lie*
*At the great foot. It moves*
*And what it moves is itself.*

Charles Tomlinson

Thus continues the cycle of degeneration and subsequent regrowth, as this rich deposit of organic matter settles and decomposes and becomes fertile, once again to feed the roots and through them the trees from which it has originated.

INTERLUDE

# *Among the trees*

Let's take a break to wander through the woods, something people do instinctively, without wondering why – although it has recently been proved that being in a green, leafy environment reduces stress.* And woods are relaxing in a way that forests are not: less threatening, more companionable, more restorative. Woodland is associated with dappled sunlight and winding paths, a mixture of trees giving different textures and colours, allowing space for the light that is excluded by a forest's dense canopy. To be lost in a wood may be inconvenient, but it is rarely frightening; to be lost in the forest is the stuff of nightmares. When Snow White runs through the forest to escape her killer, the trees too become evil, with leering faces and claws reaching out to catch her – something that could never happen in a wood, more likely the haunt of elves and fairies.

The so-called fairytales as told by the Brothers Grimm – and how grim they are – rely heavily on forests for their atmosphere. Little Red Riding Hood, sent off into the forest and coming to a sticky end inside the wolf;* or Hansel and Gretel, abandoned in the forest and seduced by the wiles of the wicked witch (but having such triumphant revenge). Not a fairytale but a ballad, 'The Babes in the Wood' appropriately has a different feel, for here the children, also abandoned, are looked after by a guardian fairy and kept alive by eating woodland berries until they finally succumb to their fate, when their bodies are covered with leaves by the birds.

This benign quality of woodland is partly attributable to the fact that it is usually deciduous, with perhaps just a smattering of conifers providing variety. Deciduous trees are constantly changing, marking spring and autumn with new growth and decay, the woodland floor bursting with fecundity and the rich smells of bluebells and leaf mould, the trees providing shade in summer, protection in winter and a sense of continuity in their annual cycles.

In contrast, we tend to qualify the word forest with adjectives such as 'ancient', 'deep', 'dark', 'haunted'. This forbidding aspect of forests was

**The ordered chaos of old broadleaf woodland: Beaulieu Wood, Monmouthshire.**

captured by Romantic artists like Caspar David Friedrich, who painted little figures dwarfed by the towering trunks and dense leafiness. In Britain, forests were the prerogative of royalty, fiercely guarded and out of bounds to the common people whose habitat and livelihood was the greenwood. A forest is not only bigger, it is more *weighty* than a wood, it has more gravitas. As an ocean is to a sea, or a cathedral to a church perhaps.

Early pagan worship took place in the forest, and the medieval cathedral builders were clearly inspired by trees: one has only to stand among the soaring trunks of a pine forest, their branches reaching over in a natural vault, to see where the idea for the Gothic nave came from. Architects ever since have copied the structure and form of trees in stone and cement, as well as using them to build with. 'Do you want to know where I found my model?' asked Gaudí, whose organic buildings spring treelike from the ground. 'An upright tree; it bears its branches and these, in turn, their twigs and these, in turn, the leaves. And every individual part has been growing harmoniously, magnificently, ever since God the artist created it.'

**A colonnade in Gaudí's Park Güell, Barcelona.**

One of the most influential books on forestry was John Evelyn's *Sylva, or A Discourse of Forest-Trees, and the Propagation of Timber,* which he presented as a paper to the Royal Society in 1662; it was published two years later, and in many subsequent editions. Evelyn was a keen gardener known for his expertise with fruit trees, but here he turned his attention to the woodland that was fast disappearing in an orgy of felling and destruction following the abolition of the royal forests during the Civil War. Drawing on his first-hand horticultural experience, he gave detailed practical advice on the cultivation and husbandry of trees, on grafting and pruning, pollarding and coppicing, in the hope of encouraging both king and people to replant the devastated woodland and thus ensure supplies of wood for the huge demands of Naval shipbuilding as well as domestic use.

A century later, the great naturalist and observer Gilbert White made world famous the countryside surrounding his small parish of Selborne, in Hampshire, where he was born and died. Although his chief interest was in the birds and animals that inhabited them, he describes the forest of Wolmer (now Woolmer), an ancient royal forest 'of which three-fifths perhaps lie in this parish', much of it not wooded at all but barren heathland, and the adjacent forest, Ayles (now Alice) Holt, renowned for the oak that flourished on its much richer soil. He records the felling taking place: 'A very large fall of timber, consisting of about one thousand oaks, has been cut this spring (viz., 1784) in the Holt forest,' which resulted in a squabble for the wood between 'the grantee, Lord Stawel. . . . [who] lays claim also to the lop and top' and the poor of the local parishes who claim it is theirs and 'assembling in a riotous manner, have actually taken it all away.' Lop and top was an ancient right of the people to take the branches from timber trees for firewood.

As well as dedicated naturalists such as these, artists and writers and even the odd composer have drawn inspiration from the woods throughout history.* In the mid-nineteenth century, shortly after Friedrich had painted his dark forests, Henry David Thoreau was making off for the woods. As he famously declared, 'I went to the woods because I wished to live deliberately, to front only the essential facts of life, and see if I could not learn what it had to teach, and not, when I came to die, discover that I had not lived.' Not so far from Thoreau's Walden, in eastern Massachusetts and half a century later, Robert Frost was writing the poems about trees and woods for which he is perhaps best remembered, such as 'The Road Not Taken' and 'Stopping by Woods on a Snowy Evening'. In England at the beginning of the nineteenth century John Clare, the son of a farm worker, was composing earthy poems based on his knowledge of the land, while the Romantic poets such as Wordsworth were wandering the hills and inevitably bringing mention of trees into their pastoral idylls.

---

Definitions are tricky: even the word 'tree' has no exact botanical meaning, and there is no strictly definable distinction between woodland and forest, only the notion of age and size. There is however, as we have seen, one of atmosphere, vaguer and hard to pin down but irrefutable. Just consider the extremes of forest type worldwide: the vast reaches of the taiga, the mainly coniferous forest that sweeps across the northern latitudes with temperatures as low as -54°C, home to moose, elk, beavers, bison, bears, lynx and tigers, in contrast to the tropical rainforest where temperatures hover around 25°C and the average annual rainfall is about 200 centimetres, each layer of this steamy habitat seething with exotic plants and wildlife. In comparison our temperate British woodland – even when it claims to be forest – may seem tame. But it too has distinctive features, if more subtle and 'temperate', in its diversity of trees and their associated flora and fauna.

On a much smaller scale, the character of our forests is also defined by geology and climate. The species found in the remnants of the old Caledonian forest, for example, a vast area once made up almost exclusivley of Scots pine but now including birch, rowan and juniper, create an ecosystem – and therefore an atmosphere – very different from let's say the New Forest in southwest England, consisting as it does of mainly beech and oak and supporting a correspondingly different list of dependants. But there is something more. Woodland in general, and forests in particular, have what may be described loosely as character. Without getting too airy fairy, it's as if they have their own spirit – and this is not unique to woods, of course, for the same could be said of moorland or heath or landscape. But perhaps due to the enclosed nature of woodland, its unique character is very obvious, almost palpable: we are surrounded by it.

Of Britain's forests, a few are known even to the least adventurous explorer through their historical associations. Sherwood, of course, legendary home to Robin Hood (though his forest was probably in Barnsdale, some eighty miles to the north); the Forest of Dean, from where Dick Whittington set forth for London with his cat and spotted handkerchief; the New Forest, scene of the deaths of Prince Richard and William II within about twenty years of each other, both while hunting, whether accidentally or not. (The so-called New Forest was mentioned in the Domesday book in 1086, taken over a few years earlier by William I for the royal hunt, the villages and farms within it being largely evacuated.)

We have what I would call quirky woods, too: full of character, very British. Wistman's Wood is now little more than a series of copses perched among the granite rocks on the upland reaches of Dartmoor high above the valley of the river Dart, an ecological and botanical wonder. For these trees are almost solely our *Quercus robur*, English oak, proud national emblem but here as stunted and wizened as petrified bonsais, no taller than five metres and enveloped in a mantle of moss and lichen. They seem, writes John Fowles, 'to be writhing, convulsed, each its own Laocoön, caught and frozen in some fanatically private struggle for existence.'

An oak 'Laocoön' in Wistman's Wood.

Similarly odd, similarly ancient, though with a very different feel is Kingley Vale, also perched on a hillside, overlooking Chichester Harbour on the Sussex Downs and composed of ancient yew trees. This spectacular site was once an Iron Age hillfort, though Neolithic flint tools and Bronze Age burial mounds show that it had been peopled much earlier. So it is hardly surprising that the wood too seems peopled; the yew, always associated with death, here spreads its many trunks in a huge canopy of dark foliage in which lurk the spirits of Druids, of Danes and Vikings and the brave men of Chichester who were slain here. Local legend has it that the ground runs red with blood on late summer evenings, but this may be no more than the berries that litter the white chalk beneath the tree trunks.

Thinking about the trees and woods that have marked my own life confirms this idea of character. The Oxfordshire beechwoods, companions of my childhood; the forest of pines and larches near Pluscarden in the northeast of Scotland, wild and deserted then, to which I escaped on horseback for relief from domestic turmoil; the trees of the London parks, each one intimately known by me and my dog; the stands of holm and cork oak, aliens at first, that I came to love during my decade in southern Spain; and now once more the familiar beeches and oaks, this time in Brittany, growing tall out of a chaos of moss-covered granite boulders. All different, all special in their own ways.

Although I cannot subscribe to the idea of trees being sentient, nor am I normally a tree hugger, I do recall a time when I went repeatedly to a tree, an old hornbeam in Hertfordshire, and rested my buzzing head against its trunk and found in it a quality that soothed my anguish. It is to do with a tree's constancy perhaps, its immovability despite constant movement. As Ted Hughes put it, 'A tree lets what happens to it happen,' something we humans can usefully learn.

# FLOWER

Few trees have spectacular flowers, or are known foremost for their flowers. The magnolia perhaps, especially when planted en masse, or the cherry's frilly tutu heralding warmer weather, or the horse chestnut whose candelabra light up those spring days when the sun we crave is lacking. The flowers of the tulip tree, which one might expect to have some impact, are poor relations to their gaudy garden namesakes, the tree's beauty lying in its overall shape and majesty. Indeed many trees flower without our even noticing it, and it is not until the berries come, or the keys, or the cones (though that is a different story) that we wake up. Think of the holly, the spindle tree, the sycamore. Tree flowers tend to be self-effacing, their fruit- and seed-bearing capacities more important than their own finery.

**Magnolia flowers.**

Like all plants, trees have an active sex life, with many variations on the basic theme of male and female. The open receptivity of the majority of flowers, their seductive scents and colours, make us instinctively identify them as female, just as the upright thrust of the stamen is unmistakably male. But their ménages and proclivities make the male-female coupling of mammals look positively staid in comparison. Trees can be unisexual, having one set of flowers, either male or female; bisexual (hermaphroditic), with flowers having both stamens and pistils; or even, delightfully, polygamous, bearing male, female *and* bisexual flowers. As if that isn't enough, some bisexual trees such as the pecan bear flowers whose sexual maturity, that is male pollen release and female receptivity, does not coincide. Just imagine the complications if all this were the norm for humans.

The differences are most often governed by whether pollination takes place through the action of wind or of insects, and although this isn't

**Each of the four ancient walls within the city of Lucca is planted with a different species of tree, here the magnolias.**

the sort of book to go into the complexities of plant reproduction – and complex it is – one can't avoid taking a quick look at pollination. Pollen, which is borne on the male stamen, is the carrier of sperm (not the sperm itself). When this microscopically fine powder lands on the female, whether flower pistil or female cone, the sperm is released, germination takes place and fertilisation follows.

The shape, size, colour and positioning of flowers is determined by what its pollinators are, as is the form and quantity of the pollen they produce. For obvious reasons, flowers pollinated by insects will be larger, flashier and more 'attractive' than those pollinated by wind or water need to be. The diversity and ingenuity with which insects are lured – bribed even – to play their part in this elaborate ceremony are a source of wonder. Tree flowers, being on the whole more modest than those of plants, tend to be wind pollinated, something for which the trees' size, tendency to sway and distribution of branches clearly suits them. Notable exceptions among large trees are the lime, whose sweet-smelling flowers are irrestible to bees, horse chestnut with its showy flowers, the tulip tree, and most of the fruit trees.

Bringer of life to the plant or tree, pollen brings misery to human beings – or those unlucky enough to suffer from hay fever. As its name implies, this allergic reaction is most often triggered by grasses, but tree flowers account for up to twenty per cent of so-called hay fever allergies. The birch is the biggest culprit, but oak, elm, pine, hazel and yew are all possible triggers, and all of them produce wind-borne pollens.

But rather than getting involved in the intricacies of plant reproduction, let's go in search of some tree flowers.

The flowers of the Judas tree – and these are truly spectacular – burst directly out of the trunk and main branches, as well as growing more conventionally from the twigs. They are pea-shaped flowers of a shocking purplish-pink clustered in dense masses (see illustration page 35), and since they appear before the leaves they totally dominate the tree during the flowering season, making it an eye-catcher against any background. It seems inappropriate that this beguiling tree should bear such a sinister name, though this may derive from its native site in the Judean hills rather than the more common association of Judas Iscariot having hanged himself from it.

**The famed jacarandas of Pretoria have earned their reprieve.**

Another tree whose flowers make it stand out from the green crowd is the mimosa. This too is a legume, member of the huge family of Leguminosae, though its bright yellow puffball flowers have nothing of the pea flower about them. Flowering as it does so early, the first glimpse of spring long before winter has departed, makes the mimosa particularly welcome when it strays far north from its native South America and adopted Mediterranean. Fluffy and innocent its flowers may be, but the mimosa tree is fiercely invasive as well as bearing fiendish spines, and its leaves curl up on being touched, lesser versions of the Venus flytrap.

One tree that is certainly known for its flowers is the jacaranda. This outstanding tree also originated in South America but has made itself very much at home in Europe wherever frost allows. Exceptionally graceful, it can grow up to 30 metres high and is covered with a blue-purple haze that dominates its surroundings in spring, and often again in autumn. The jacaranda grows well in towns, though it too is invasive as well as thirsty. Pretoria, known as 'Jacaranda City' for the hundreds of trees that line its streets and adorn its parks, recently underwent a campaign to have them removed, but the outcry was such that most of them remain.

Northern trees are more sober in their flowering. Flowers tend to be small and pale, most often white. Of native British trees, the most exuberant are perhaps the thorn trees – often not much more than shrubs – whose spiny branches burst into a froth of springtime blossom. First of these is the blackthorn, a small hedgerow tree that flowers before the leaves come, delineating the fields with white. Its fruit, the sloe, delicately dusted with white over a purple-black skin, is so bitter that to bite it turns one's mouth inside-out.

Blackthorn is followed by the hawthorn, fuller of flower and mellower of fruit. Known as 'sister flowers' though not related, both are associated with fertility and are rich in symbolism and legend. The blackthorn represents good and evil, its pure white blossom contrasting with the black spines, said to have been used by witches to inflict injury. The hawthorn, a member of the rose family whose flowers may be white, pink or red, played a major part in pagan Mayday rituals and is held to have many healing properties – sedative, anti-spasmodic and diuretic.

As dramatic in its opposites as blackthorn, though more stark, is the almond tree, one of the first to blossom in spring, whose 'bare iron hooks' break forth in 'flakes of rose-pale snow' in D.H. Lawrence's poem, 'Almond Blossom'.

Though it grows reluctantly in Britain, the almond is too far from its native eastern Mediterranean to thrive here, and those that survive seldom become the black, twisted sculptural forms celebrated by

**Almond tree blossom in February, Sierra de Aracena, Andalusia.**

Lawrence. Happier in the balmy climate of California, almonds are now grown there commercially in such staggering quantities that at blossom time an army of professional beekeepers from all over the United States descends on the immense orchards with refrigerated pantechnicons carrying the bees needed for pollination. This has, not surprisingly, led to all sorts of problems for both bees and trees.

Some of the most rewarding flowers of our British trees are found in the orchard. Those of the apple prove their place as members of the sprawling rose family, being pale pink, delicate, open and sweetly scented. Not strictly native, apple trees were introduced to Britain during the Roman occupation, along with pear trees. Pear flowers, less beautiful individually than those of the apple, gain their effect from the wild abundance with which they smother the tree in a blanket of white.

The ornamental varieties of orchard fruits, notably the cherry and plum, have been bred expressly for their flowers, but in the process have been made to sacrifice their fruitfulness.

Though catkins may be suggestive of male sexual energy they are in fact made up of tiny flowers. Whether erect or dangling, smooth or hairy, catkins pulsate with energy. Hazels are the first to produce their catkins, long and hanging loosely, like lambs' tails, closely followed by the willow whose grey furry stubs are said to resemble kittens' paws, hence the name pussy willow. The Chinese, for whom willow catkins symbolise the coming of prosperity, use the branches in celebrating their New Year, the Spring Festival. In ancient Greece, sprays of pussy willow were laid on the beds of infertile women to work their magic.

Stamens in the catkins of the mulberry tree are energetic indeed, holding the record for the fastest movement in the entire plant kingdom. Acting as catapults, the stamens eject the pollen at an astonishing 560kph, half the speed of sound. Knowing this makes it less surprising to learn that the Druids used pollen as a primitive type of firework, throwing it in handfuls on to ceremonial fires where according to the type it sizzled, crackled, flashed or exploded, giving off different colours as it went.

## Not the Same Tree

The sweet or Spanish chestnut comes alive – 'blossoms' – with its catkins, butter-yellow spikes that cover the tree with a glow that vibrates the hillsides in early summer, well after the leaves have appeared and when most other trees have settled into greenery. The catkins, as much as 20 centimetres long and upright, contain both male flowers (above) and female beneath them. Their transformation within the space of a few months into dark, shining, nutritious nuts inside a viciously spiked shell is a marvel indeed.

Although cones are clearly not flowers, they perform the same function – namely bearing pollen (male, sperm producing) and ovules (female), usually on the same tree. Cones are actually made of modified leaves, and although some are pendulous and some upright, some scaly and some smooth, compared with flowers they are fairly uniform. Pollen from cones is mostly wind borne. Usually the cones open when ripe to release the seed, though some have to be broken open by birds – and certain pine cones are even stuck together with a resin that has to be melted by fire, hence the part that forest fires play in natural reproduction.

**Male chestnut catkins dominate; the small female flowers are hidden at their base.**

**Male and female pine cones in various stages of development.**

Though the tree flowers' main task, their purpose, is to bear seed and fruit and thus propagate themselves, man has come up with many other uses for them. Flowers and herbs were the earliest known medicines, and the flowers of trees – as well as their bark, leaves and berries – played, and still play, a significant part in the materia medica of natural remedies. For ailments physical, from the common cold (elderflower) to acne (hawthorn), and emotional, from grief (ashoka, the 'Sorrowless Tree') to hysteria (lime tree), there is an ancient, proven, tree-flower remedy.

Most substances used in natural cures have a homeopathic version which often, in the way of homeopathic healing, either reflects or opposes – sometimes both – the qualities of the host, whether animal, mineral or plant. *Vitex agnus castus*, the so-called chaste tree, is most often regarded as being sexually stimulating, and its flowers, leaves and berries are used in homeopathy as Agnus castus to treat associated disorders of the menopause and impotence. The male flowers of the Chinese sumac or tree of heaven – male and female flowers are on different trees – are not as pleasant as they sound, smelling strongly of cat's urine and being mildly toxic; the seed pods, red with yellow bumps, look like an illustration from a medical handbook on skin diseases, and the remedy Ailanthus is aptly used to treat typhoid, scarlet fever and skin eruptions in general.

Bach flower remedies focus entirely on the emotional and mental states which lie behind physical illness. Of the 38 remedies, 17 are made from tree flowers. Similar to, but not the same as homeopathic remedies (Edward Bach was a trained homeopath), they work on a vibrational level which also reflects that of the host flower. In this context I particularly like the symptoms listed under the crab apple: self-hatred, obsessiveness, over-anxiousness, fussiness – in short, crabbiness. You can just picture that tight mouth pursed around the bitter little apple.

Both homeopathy and the Bach remedies are regularly trashed by scientists who, despite their oft-quoted 'rigorous clinical trials', cannot find out how they work. But that they do is undisputed by their thousands of practitioners and devotees worldwide.

Elderflowers not only smell heady and erotic, they make a wine that stands up well amongst home-grown wines, even if these are derided by wine snobs. To my mind elderflower wine has the qualities of a Gewürztraminer – flowery and aromatic, with an edge to it. But if you don't like the smell of elderflowers, you won't like the taste of the wine.

Thoughts of the palate and of sweet smells lead us to another of the tree flower's gifts, honey. Though most honey on the supermarket shelf is a blend made from different flowers, you may also find there pure honeys such as acacia, lime and orange blossom – though none of these compare to the honey made by your neighbour with his few well loved and tended hives, scraped fresh out of the comb with flakes of wax still sticking to it. Nectar, liquor of the gods.

Nectar – not to be confused, though it often is, with ambrosia, food of the gods – is a sugary liquid produced by flowers as a lure to

**Lime flowers may look insignificant but are rich in perfume and nectar.**

pollinators, principally bees and butterflies but also moths and, in warm climates, hummingbirds. It is this sugary bribe that provides the sweetness of honey, in the form of fructose, glucose and sucrose. (A different sort of nectar is produced on different parts of the plant, on leaves, fruits or stems, to attract predatory insects which will defend the plants from invaders.)

Honey has different smells and flavours according to its flower source, some of these very subtle. The smell of the flowers themselves opens up a Pandora's box of associations and memories: of childhood, of foreign holidays, of summer evenings, of mountains and heaths and hillsides, or just of the local park. Lime trees spring to mind as I write – nothing to do with the green citrus fruit, of course, but the huge, gracious trees laden with tiny, sweet-smelling flowers, the common lime or linden tree, whose genus is prettily named *Tilia*. Some people curse lime trees for the 'honeydew' that drops from them in late spring, a sticky sap produced by the tree to attract ants and aphids that will also coat anything left underneath its branches – particularly the cars of suburban man, who then complains to the local council and tries to get the trees removed. ('Suburbia is where the developer bulldozes out the trees, then names the streets after them,' remarked Bill Vaughan.*)

Besides being the source of some of the most prized honey, pale and distinctively flavoured, lime flowers when dried make a gentle relaxing tisane, particularly popular in France. The flowers contain flavonoids, the antioxidants currently fashionable in dietary supplements,* and are used in herbal medicine for a variety of mild disorders.

The natural scent of tree flowers wafts over us free and unsolicited, but the flowers of several species are used for making commercial perfumes and essential oils that sell for often ridiculously inflated prices. In the Far East trees are grown specially for the purpose. The ylang-ylang, a native of Indonesia also known as the perfume tree, has yellow or cream spiked flowers whose smell prompts descriptions worthy of the most dedicated wine buff: 'rich and deep with notes of rubber and custard' – no wonder it is also considered to be an aphrodisiac. Ylang-ylang oil, obtained from the flowers, is a component of macassar oil, which gentlemen in the nineteenth century used to keep their hair slicked back (and which led to that most Victorian of objects, the antimacassar, usually – like the doily – crocheted, but now, on the back of seats in planes or railway carriages, made of something nastily synthetic).

**One of the many delicate shapes and shades of ylang-ylang flowers.**

A tree whose botanical name is confusingly both *Michelia* and *Magnolia champaca* is known as the Joy perfume tree, which gave that name – or maybe it was the other way around – to a perfume made from its flowers in 1929 for the French couturier Jean Patou. This expensive and famous perfume, which outdid its rival Chanel No.5 as 'Scent of the Century' in 2000, also contains ylang-ylang, though its main ingredient is jasmine. Most perfumes are a blend of different scents, and another famous one, Dior's J'Adore, includes flowers from the Joy perfume tree. Acacia, orange blossom, mock orange blossom (syringa) all offer their distinctive smells to scents, colognes and essential oils.

But now let's leave the cloying, claustrophobic aisles of the department store and get back outside to the trees as their flowers perform their natural function of producing fruit and seeds.

Opposite: **Lime trees come into flower late but with striking ebullience.**

# FRUIT

Fruit – what lush, juicy, juice-inducing images the word conjures up. Just thinking about biting into a ripe pear – that scene in the film of *Tom Jones*, do you remember? – or a greengage warm off the tree, a mulberry spurting its blood down one's front, or a cherry – black, white or red, how can one choose – makes the saliva flow, along with the associated memories. Ted Hughes describes how as a young man in post-war London he bought a peach 'from a stall near Charing Cross Station. / It was the first fresh peach I had ever tasted. / I could hardly believe how delicious.' As a war baby myself, raised on puréed swede because there were no oranges available to provide vitamin C, I can identify with this incredulity. Once a year, at Christmas, we received from relatives in Jamaica a box, made of blond wood and full to the brim with fine white sugar. Inside, like prizes in a bran tub, were crystallised fruits – exotic things, unknown to us: guavas, mangoes, pineapple, figs . . . I can see them, feel the excitement now as my brother and I dug and delved.

Supermarket fruit, picked too early, frozen and transported half way across the world, may often disappoint (those perfectly matched, uniform apples that taste of, well, of nothing), but citrus fruits travel well and have improved immeasurably in the last half century. Take grapefruit, which I remember being presented, rather daringly, on the breakfast table in the 1950s: wizened, greyish-green and sour, with little juice and a tough inner membrane, halved and cut into segments, a cherry in the middle and scattered with sugar in a vain attempt to make it appealing. How different are the plump, pink-fleshed Californian beauties we now enjoy, if guiltily: sweet, full of juice and best eaten alone – if Tom Jones isn't around – cut in half and held in both hands, over the kitchen sink to catch the drips. Not even in the interests of the environment am I prepared to swap this for puréed swede.

Citrus trees have much in common. Orange, lemon and grapefruit all have a distinctive feel, a thin green bark that is more like a skin and if scratched smells like the fruit itself, and flowers that smell divine. All the citruses are rich in vitamin C; all flourish in warm climates and few can

Opposite: **Peaches 'into my hands themselves do reach', wrote Andrew Marvell.**

cope with frost, which perhaps justifies us northerners importing them. The trees can be grown in northern Europe if protected from frost, and they often are, but as ornamentals for the beauty of their shining leaves and sweet-smelling flowers; the fruits seldom make it to the table.

One of the joys of my time spent in Andalusia was having an orange tree in the garden. (I was going to say 'owning an orange tree', but I got used to visitors helping themselves to fruit without asking, and one day came out to find a stranger filling his bag. When I said indignantly that it was my tree, he looked at me as if I had said something stupid – which of course, when I came to think about it later, I had.) These oranges were small, sweet, tangy, perfect for juicing, and lasted from late November to early April, a winter of tiny golden suns and free vitamins.

**These oranges in Andalusia hold their own in a January frost.**

Seville oranges, thick-skinned, knobbly and very bitter are of course responsible for what strangely is a most British delicacy, marmalade. Exported in huge quantities from Seville and its surroundings to provide marmalade on practically everyone's breakfast table in the UK, they were unobtainable when I lived in the area; the only chance – unless you were lucky enough to have a tree on your land – was to come across swathes of branches cut down in the early spring, after they had served their purpose as ornamentals, and dumped by the road, laden with fruit. Once, when stuffing my car with this booty, I was tooted at by a passing Spaniard who shook his finger at me, not because I was doing anything wrong but to tell me they were inedible.

'Lemon tree very pretty, and the lemon flower is sweet, but the fruit of the poor lemon is impossible to eat.' But the poor lemon tree has turned its shortcoming to advantage, has turned its lemons into lemonade, and if that sickly drink has come a long way from its sharp, aromatic source, it is hard to imagine modern cookery, whether sweet or savoury, without the enlivening effect of lemon juice.

We have grown used to buying exotic fruits from all over the world, but though we may know them by name on the supermarket shelf, or even on the market stall, we probably have little idea of what their parent trees look like, how they grow and in what conditions of soil and climate. What does a mango tree look like? I confess I had to look it up – and fascinating it is, an elegant tree that can grow up to 40 metres high, with many different varieties including one developed in Cuba called *huevos de toro*, a visual-verbal delight when you see the fruit hanging on the trees. And what fruit it is – pale yellow to orange blushed with red, to lime green, to deep blue-green with a purple bloom, the flesh likened to that of a plum crossed with a melon but with a flavour all its own. Are you salivating yet? I also learned why mangoes were made into chutney in the days of pre-refrigerated transport, and from that how the word 'mango', derived from the Tamil, came to be used as a verb, to pickle.

Such are the avenues of virtual delight down which one can wander in the twenty-first century. But seductive though they are, the aim of this book is to put us in touch with – literally perhaps – the *reality* of trees, their smell and feel and earthiness. So let us go back to the temperate climate of Britain, to the fruit-bearing trees of the gardens, hedgerows and orchards that still cling on in a hostile, pest-controlling, plastic-wrapped environment.

Many suburban gardens boast at least one fruit tree, and most usually that is an apple. Apple trees do particularly well in our climate, although like so many other fruit trees they originally came from Asia. The trees are small and relatively pest-free, the fruit doesn't depend on sun to ripen, and the varieties are legion – more than seven and a half thousand, we are told. And of course, an apple a day . . .

The apple was probably the first fruit tree to be cultivated, and apples played a part in pagan and religious mythology from the Greeks to the Celts long before their key role as the forbidden fruit in the Biblical story of the fall of man. In Greek myth, the Garden of the Hesperides was an apple orchard in which grew the sacred tree whose golden apples gave immortality, gift of Gaia to Hera on her wedding to Zeus. The eleventh labour of Hercules was to bring back one of these apples, fiercely guarded by nymphs, the Hesperides, and by the serpent Ladon: serpents in paradise a persistent myth.

In Celtic legend, a silver bough (not to be confused with the golden one made famous by James Frazer) from a magic apple tree bore golden apples which hung like bells and made sweet music that lulled people into a trance, thus bridging the gap between this world and the next.

For us today the apple is more mundane, if no less important: more than 70 million tonnes are grown worldwide every year. One of the easiest fruits to store and keep, apples are ideally suited to modern

**The apple as temptation: from an early German Life of Adam and Eve.***

**Fruit trees can be manipulated into many different shapes and sizes . . .**

methods of refrigeration and transport; they travel well, even if in doing so they inevitably lose some of that delicious shock of a fruit newly picked from the tree. In fact, I find most apples one buys in the shops, even in street markets, are rather like High Street curries – uniform and instantly forgettable. Fruit of the gods they no longer are, though there are still exceptions.

**. . . but are happier left to sprawl.**

In his book *The Tree*, John Fowles writes movingly about his father, a withdrawn and rather sad figure who found solace for his unfulfilling life in growing fruit trees. For these were 'more than trees, their names and habits and characters on an emotional parity with those of family'. The young Fowles, having learned at his father's feet the art and discipline of proper pruning, took the other side in adulthood, allowing – encouraging even – his own much loved garden to revert to wilderness, leaving it to 'my co-tenants, its wild birds and beasts, its plants and insects.' But even from this perspective over the fence that divided them, he retained an affection for the father who had taken such active pleasure in the trees that he pruned, cosseted and prayed for, and whose fruits he treasured.

All of the delightfully named Victorian varieties grown by the father Fowles – Charles Ross, Lady Sudeley, Peasgood's Nonsuch, King of the Pippins – are availabe from specialist nurseries today, but you won't come across them in the supermarket. This is because they *don't* travel well, nor do they keep. Fowles's father 'knew exactly when they should be eaten. A Comice pear may take many weeks to ripen in store, but it is at its peak for only a day. Perfection in the Grieve [apple] is almost as transient.'

Pears, yes. How many times have you bought them, hard and green, squeezing them gently every day until they seem to be ripe – only to miss the moment and have them fall apart in a tasteless mush. There are now honourable exceptions to this, carefully bred to behave well, and it amused me to speculate whether the Conference, one of the most reliable, was dreamed up and named by an earnest gathering of fruit growers, or importers in business suits maybe, seated round a table. (I wasn't so far out. It was named after the British National Pear Conference, where it was first exhibited in 1895.)

Pears, even those grown at home, can be very dull. We had several big old pear trees in Andalusia which most years dripped with pears, but they were so tasteless whatever you did with them – cooking them in wine, spices or honey had no effect on their stubborn blandness and gritty consistency – that the fruit the donkey didn't eat (and she wasn't very impressed either) ended up as a thick gunge circling the base of each tree that finally disappeared only with the winter's frost.

The French have come up with some ingenious ways of making pears interesting, the most bizarre of which involves placing a bottle over the developing fruit so that you end up with not a ship, but a fully grown pear in a bottle. Preserved in alcohol, it may look disturbingly like a specimen foetus, but if you can overcome your misgivings you will enjoy both liquor and pear, each imbued with the other's flavour.

An eau-de-vie made from pears, called Poire William, is made in Alsace and Switzerland; it too often comes with a pear in the bottle and smells deliciously of the essence of pear, but the taste seldom lives up to its promise.

And now a short diversion to acknowledge that, botanically speaking, neither apples nor pears are fruit, nor are any members of the large Prunus family which along with plums includes peaches, nectarines, apricots, cherries and, surprisingly, almonds. Technically speaking, apples and pears are pomes, as are the 'berries' of the rowan and hawthorn; that beautiful fruit the quince is a pome too. You can tell a pome by the remains of the flower at the opposite end of the fruit to the stalk, and by its pips. In contrast, the Prunus family are all drupes, which have softer flesh than the pomes and a stone or pit inside which lies the seed, or kernel, metaphor for the essence of the matter. Other drupes are avocados, olives, dates and mangoes, the latter's stone being remarkably like a cuttlebone in shape and colour.

**Beware the seductive appearance of this drupe, the olive - hard, bitter and inedible until laboriously treated, see page 100.**

Plums are a slightly neglected fruit, maybe because they really *do* need to be eaten straight off the tree. Picked before ripe, refrigerated and stored, they lose most of their flavour, becoming deadened and dulled. The uniformly large, round, imported ones that may seduce you with their colours, most usually bright yellow or almost black, are only edible to my mind if stewed. But anyone who has ever picked a plum or a greengage, warm from the sun, still covered in a silvery bloom, bursting with juice and flavour, will know that there is little in the fruit world to compete with it.

In England we tend to think of plums as oblong and red or golden, and most people would be pushed to name more than the ubiquitous Victoria. The French are more adventurous, with a wide range of common smaller varieties with enticing names like mirabelle and quetsche. But once you start to look, there are exotically named English varieties too – take for example Dittisham Ploughman, Angelina Burdett, Warwickshire Drooper, Thames Cross, which make one want immediately to rush to the catalogues – or Google – and discover their histories even if one never gets the chance to taste such a rarity.

Plums grow well in the wild. Some may be escapees from the orchard, sprung from a stone dropped by a bird who enjoyed the fruit and passed it on, as birds do. But many are genuinely wild. Closest to the cultivated damson, wild plums – also called bullaces – come in colours from darkest blue-black through red, crimson and orange to pale yellow, but all have thin skins and yellowish flesh. They also boast one of the highest food values of any other fruit, 20 per cent carbohydrate, as well as a load of vitamins and minerals – all free in the hedgerows. And all too make wonderful jam.

Though other members of the plum family such as peaches, nectarines and apricots grow more happily in warmer climates than ours, they can be induced to fruit if protected from frost, though the fruit often lacks the fullness of flavour bestowed by hot sun. But I recall, like Ted Hughes, the taste of my first nectarine, white-fleshed, sweet and dripping with juice, picked from an old espaliered tree in a walled garden in South Devon and surpassing any import I have tasted since.

Unlike their sturdier relatives the plums, nectarines and peaches have specific and rather limiting climatic requirements, for while the flowers cannot tolerate frost, the tree needs a period of chilling in order to initiate its flowering. In their native habitat this period can be as long as forty days, though strains have been developed that need only a quarter of that time. Then of course the fruit needs hot sun to ripen in as well, all of which makes it clear why growing nectarines and peaches successfully in temperate climates is not an easy task.

Although everyone is happy to call the fig a fruit, technically it isn't one either. Unique in its way of growing straight out of the trunk like a bud, it is a 'false' or multiple fruit formed from a cluster of tiny flowers. Fig trees are resilient and will survive northern winters, though they would prefer to thrust their roots into a baked Greek hillside, and figs too struggle to reach perfection without hot sun.

**Figs ripening in Andalusia, still green but showing promise.**

I was surprised to learn that the fig comes from the same family as the mulberry, the Moraceae, particularly as the mulberry is wind pollinated whereas the fig is dependent on a particular wasp for its pollination. The mulberry is not a berry but a false fruit too, formed in the same way from tiny flowers. The connection between the two is graphically described as 'the fig correspond[ing] in structure to an invaginated or inside-out mulberry'. Sexual imagery comes naturally when describing figs, these 'self-conscious secret fruits' as D.H. Lawrence calls them: their display when cut open, or bitten into, is shameless, and it is not by chance that the Spanish word for fig, *higo*, is close to that of the female genitalia, *higa*. But whatever its associations, seeing a tiny fig sprout as if by magic from the bare branch, watching it slowly plump out, turning from green to purple (or yellowish-white), then picking the first one, biting into its strange, almost spongy flesh – this is one of those joys of fruiting trees that seem God-given.

Fruit-tree cultivation is a complex and highly skilled process, as well as a lucrative one, none of which has a place here. Grafting, rootstocks, self-pollinators and the like are best left to the specialist. Trees of the twenty-first century, commercially grown in head-spinning quantities, bred low for easy picking, pruned and primped like poodles in a parlour, are a long way from their sprawling wayward ancestors. The list of fruits is long too, and growing, and tempting to follow in its diversity and richness of colour, texture, flavour and nutritional value.

But let's take a look at those 'fruits' that hardly seem to deserve the name, so far are they from the heading of lush and juicy. One of the most rewarding of these is the quince – rewarding once it is cooked, that is. The beautiful sweet-smelling pink flowers of the quince tree turn into a fruit – a pome – like a large pear, golden yellow when ripe and covered in fine white down. Inside, quinces are brick hard and gritty and often filled with bugs, particularly earwigs. Cutting them up is hard work, and when you have them boiling in a pan like a load of grey washing, you may wonder why you bothered. But slowly, and more quickly once the sugar is added, an alchemy takes place and the base quince is turned into gold – or more accurately into amber – and a divine smell envelops the kitchen. (The word marmalade comes from the Portuguese for quince, *marmelo*.)

Another fruit, this time a drupe, that needs working at before it is edible is the olive. If you have ever picked one from the tree and bitten into it – as most people not brought up in olive country have done, once and only once – you will wonder who first discovered that it *could* be made edible. It's a long-drawn-out process involving much salt and daily changing of water over many days or weeks, and though as the proud guardian of four olive trees on my Spanish land I tried several times, following recipes, neighbours' advice and local lore, they always tasted disgusting and ended up on the compost. Which is maybe why olives sold commercially are 'cured' in one day, in lye. In turn, this is maybe why they are soft and squishy and taste of nothing at all, and bear no relation to a properly cured olive.

The transformation of these hard, bitter little objects into something as golden (or better, dark green), as nutritious and as delicious, as deeply satisfying on so many levels as olive oil, is one of man's happiest collaborations with nature. I cannot imagine life without olive oil. I cook everything possible in it, pour it over every salad, rub my hands with it – even, with some guilt, give it to the dogs, which is no doubt why they are so healthy. The guilt is a leftover from that wartime upbringing, when olive oil came – when it came at all – in bottles of perhaps four fluid ounces (120 millilitres), no more than phials, bought from Boots the Chemist and kept in the medicine cabinet, to be used very sparingly to clear the wax from one's ears.

The story of the olive, both tree and fruit, as food and as symbol, is long and fascinating and well documented. The first known mention of the production of olive oil is in the 6th millennium BC, near Mount Carmel in Israel. From then on the 'liquid gold', as Homer called it, is a constant in literature from the Greeks and Romans to the Old Testament, in East and West, in allegory, parable and fable.

Claims for the fruit of the date palm go back even further, to 7000 BC, and also feature in the lore and literature of both Islam and Christianity. Dates, also drupes, have enough sweet flesh on their stones to be comfortably regarded as fruit, and are more immediately gratifying than olives. Eaten straight from the tree – or rather, the ground – sweet and plump, these are very different from the fibrous and often dry little corpses that arrive in our shops at Christmas, in their cardboard coffins complete with plastic twig to dig them out with. These too are sweet, rather unattractively so, for the sugar content of dates whether fresh or dried is as high as 80 per cent.

Olives and dates don't flourish in our British climate, but we have some hard-fleshed fruits of our own, mostly wild since their use is limited and they cannot now compete with the imported exotica. Richard Mabey, in his excellent *Food for Free*, gives ideas for how to use these fruits, such as the medlar that must be bletted – allowed to become half-rotten – before it is eaten. Sorbs, which come from *Sorbus torminalis*, the wild service tree, are like small hard pears, and they too need bletting. Other Sorbus species, such as the rowan (mountain ash) and whitebeam, all from the rose family, bear edible fruits that to the layman look like berries, but all of this lot are in fact pomes, with the telltale whiskers opposite the stalk end of the fruit.

Berries. Where do they fit in? Not, surely, into our definition of fruit, though 'soft fruit' is the collective name for the undisputable berries of the British garden, the raspberries and currants and strawberries. But if, as I do, you think of a berry as being small and often hard, you may be surprised to learn that pumpkins and water melons are berries, as are the tree-borne persimmon and avocado. In fact botanically speaking the citrus fruits are berries too, as are bananas,* but that's getting a bit technical. Botanical fruits may not be culinary fruits, and vice-versa, and the terminology is confusing. The juniper berry, for example, which gives gin its distinctive scented flavour, is not a berry at all but a cone. So for our non-scientific approach, let's go with the cooks and treat berries as fruits.

Most berries, particularly the soft fruit, are synonymous with health, and in our diet-conscious age they play a big part in the keep-fit campaign. Packets of frozen 'red fruits' can be bought all year round, packed with vitamins and minerals and claiming to offer protection against cardiovascular disease and even cancer – their very redness proof of nutritional value. Of the tree-borne berries, one of the best to my mind is the mulberry, that ruby-red fruit whose juice is so shockingly like blood, spurting and staining in such a bloodthirsty manner. In fact they come in black and white as well as that red, all equally nutritious. The reason that mulberries aren't seen more often off the tree is presumably just because they are so volatile, which, combined with the fact that the crop ripens over a long period, must make them daunting to grow commercially.

**Persimmons decorate the tree in winter, like baubles on a Christmas tree.**

As are persimmons, though for different reasons. This beautiful fruit – I can't bring myself to call it a berry – must be on the point of collapse before it is edible. Until that moment it will leave your mouth full of dry fur, most unpleasant. Judging the right time to pick them is tricky, and many of the precious fruits from my Spanish tree landed with a splat, to the joy of dogs, donkeys and chickens who fought each other to get there first and slurp them up. But much as I love eating the fruit, with its orange jelly-like flesh and indeterminate flavour, I would have a persimmon tree for the sight of them alone, their golden-pink globes glowing like fairy lights long into winter after all the leaves have fallen.

Another fruit that surprisingly comes under the heading of berry is the pomegranate. This word, literally 'seeded apple', may have given its name to the city of Granada (the Spanish for pomegranate) and certainly gave it to the hand grenade – this for the shape, though in its extreme hardness it would surely be an effective weapon itself. Inside the orange-red shell, imbedded in white membrane, are small seeds surrounded by their casings, the arils – up to 1400 of them in each fruit. These exquisitely coloured if slightly watery seeds can be eaten just as they are, or made into juice or syrup; they are an integral part of the cuisine of Persia, where they originated, and have a host of nutritional and medicinal applications. Which perhaps makes it less surprising that this strange fruit played a key role in myth and ancient history, and in Judaism it displaces the apple as the forbidden fruit in the Garden of Eden.

Closer to home, the small wild berries so appreciated by birds are rich in vitamin C. Although none to my mind compare with blackberries, which for our purposes do not count though they thrive in spaces in woodland, whitebeam and service berries when bletted are passable to eat, while rowan and hawthorn are good as jelly. And of course from the glowing black elderberries come home-made wine and other more esoteric potions, such as the Pontack Sauce described by Richard Mabey, a thick, punch-like drink that is best kept for seven years before being drunk.

The edible fruit of the strawberry tree – a true berry – resembles a lychee inside its tough case, though it doesn't taste so delicious, in fact it doesn't taste of anything much and has a gritty texture. (Its Latin name, *Arbutus unedo*, is derived from *unum edo*, 'I eat only one' – as with the olive.) But it more than makes up for this with its decorative value. The rough-textured circular fruits hang gracefully in sprays, in varying sizes and stages of ripeness, the colours going from yellow to orange to a bright pinkish-red standing out against the evergreen leaves. And as if this weren't enough, the tree bears both fruit and next year's flowers, bell-shaped and pale cream, at the same time, a sight to gladden the heart in early winter.

# NUT

As with berries, the botanical definition of what is a nut differs from the common or culinary one in many cases, but here we will include any kernel inside a hard shell that is edible to humans, as well as the horse chestnut, which is not. With the exception of peanuts – the groundnuts notorious for costing the British taxpayer millions of pounds when a postwar scheme to grow them in what is now Tanzania failed after only five years – all of these come from trees.

Nuts are a nutritionist's dream, containing all the unsaturated fats and fatty acids that we should eat, and none of those we shouldn't (apart from coconuts and palm oil, as we shall see). They provide protein for vegetarians, and all of us with arginine, antioxidants and many minerals. Most nuts are wonderfully easy to eat as well: they keep for months, need no cooking, are not messy, and as a bonus look decorative in their bowl on the sideboard.

Perhaps the king of nuts is the walnut. Borne on beautiful, graceful trees, hardy, easy to harvest, comparatively easy to shell, full of all those things that are good for us – and delicious. Shelling a walnut is every time like performing a minor piece of surgery, as the nut cracks and the densely folded and convoluted halves are revealed in their brainlike form. Sometimes one blunders and is left with a crumbly mess of nut and shell, in which case put it aside for the birds and start again. Or give it to the dogs. My two Spanish dogs, brought up foraging under the walnut trees, learnt how to deal with the shells from an early age; now, no longer in walnut territory, they appear the moment they hear me pick up the nutcrackers.

Talking of dogs, there's a politically incorrect saying, 'A woman, a dog and a walnut tree, the more you beat them the better they be.' Certainly beating still goes on in Spain – if not the women, often enough the poor dogs. As for the walnuts, long sticks are kept specially for bashing them down, which in also breaking the long shoots is supposed to encourage the growth of shorter, fruiting spurs.

**'The royal nut of Jupiter': nutritious, aphrodisiac, an antidote to poison.**

What surprised me was the difference in their form; from a dozen or so trees I harvested nuts that were small and round, large and round, rugby-ball shaped, thin and pointed, some with such a point it was almost a spine, some fairly smooth, others deeply ridged. The consistency of the nuts differed too: softness and hardness, dryness and oiliness – and, more subtly, the flavour.

Hazelnuts have many of the qualities of walnuts, and the advantage of growing wild in hedgerows and coppices. The disadvantage of this is the competition, mainly from squirrels. Many times I have noted the position of promising trees, watched the catkins do their stuff and the flowers turn into small green nuts with those delightful spiked ruffs, only to find the tree surrounded by a litter of nuts, all of them holed. Somehow enough must remain for the various rodents' winter storage, if not for us too.

If you are a chocoholic as I am, thoughts of hazelnuts lead automatically to almonds, for though the nuts are quite different they both combine magically with chocolate. The almond is well defended; its grey-blue-

**Hazelnuts, prized food of the gods, man – and squirrels.**

**Chestnuts demand to be roasted, as here on a street in Paris, 1857.**

green hull clings until the nut is ripe, the shell is very hard, and the thin sheath encasing the kernel comes off easily only after being plunged into boiling water. Almonds are bitter-sweet and contain the deadly cyanide, but their distinctive flavour makes it all worth while.

The Italians are good with almonds. Think of that most lovely confection the macaroon, which (unlike the dried-up packeted versions on the biscuit counter) is crisp and light and gooey in the middle, to be eaten with a splash of amaretto to make it true perfection. So it was disappointing to find out that the famous Disarrono liqueur in its square-hatted bottle, first made in 1525 and described by its makers as 'smooth, velvety and full of almond aromas', is not made from almonds at all, but from apricots – the apricot of course being related to the almond in the Prunus family.

Even more difficult than the almond to extract from its shell and protective coating is the chestnut. Possibly the most beautiful of nuts, sweet chestnuts drop – seem to burst – from their spiny green case, usually twins and often triplets, shining as if varnished, irresistible to any passer-by, human or animal. They can be eaten raw, peeled of their clingy, slightly furry inner casing, but chestnuts really come into their own when roasted, preferably on a charcoal griddle, and preferably eaten in the open air, round a bonfire or outside the British Museum. When cooked until the skin is almost black, the inner coat will come away easily, though burnt fingers are a necessary part of the ritual.

Although this is how most modern Westerners think of chestnuts, as a bit of fiddly fun at Halloween and Christmas, they were for a long time much more than this to many people in the Mediterranean: the source of much-needed carbohydrate in the form of flour and a type of polenta, as well as being used for fattening up their animals. And Richard Mabey informs us that Goethe's favourite dish was chestnuts with brussels sprouts – a Germanic version of bubble and squeak.

The dubious pleasure of this dish brings me to the least palatable of the nuts, the coconut – not in fact a nut but a drupe (see page 97), and to my mind not worthy of the term culinary nut though classed as such. Sweet, insipid and bitty, the coconut deserves to be desiccated, though that doesn't add to its appeal. If not target practice in a fairground, this overgrown nut – which is wickedly high in saturated fat – is best turned into oil and used for soap and cosmetics, or made to serve as a tool or cooking implement.

A true nut on a much smaller scale is the acorn. Edible for humans though not much fun to eat either, acorns are the food of the gods to many animals. The Iberian pig, a domestic pig interbred with the wild boar, is still reared and fattened in parts of Spain exclusively on acorns from the holm and cork oaks, producing the dried ham for which these areas are famed – ham with a flavour like none other, which sells for exorbitant prices worldwide. Deer, cattle and horses also eat acorns, though some people hold that they are toxic for horses; maybe it depends on the type of acorn – or the type of horse. Mice and squirrels, pigeons and jays all love them. Some devour them on the spot, others cache them for a rainy day: acorns store well and keep their nourishing qualities, a high quantity of protein, carbohydrates and fat. And, evidence of another link in the food chain, a small clay pot with air holes was dug from the Vesuvian ashes that obliterated Herculaneum and Pompeii in AD79; it had a cup at each end for water and acorns, and would have been used for fattening doormice, a Roman delicacy.

Acorns are sometimes called oak mast, a word that usually applies to the nuts of beeches. The beeches don't start to produce a good crop of nuts until they are about fifty years old, when they scatter large quantities of prickly, star-shaped husks that spring open to release the nutshells, three-cornered and shiny and devilishly difficult to open in order finally to get at the small kernels. If you have the patience, the nuts are edible but not exciting, nutritious but not particularly so, and there is a high proportion of duds so you have to be very patient – or very hungry – to bother; otherwise I suggest you leave them for the mice and squirrels. (Though in times of plenty it is easy to be dismissive, and the lives of many a starving peasant family and fleeing partisan have been saved by beech and oak mast.)

Quite why a stale joke is an old chestnut I don't know, but it probably applies to the horse chestnut, non-edible but almost as beautiful as its Spanish counterpart and just as much fun to pick up – especially if you are young and out to conquer. And that's a bad joke too, as the game of conkers actually takes its name from the snail shell (conch) it was originally played with. A vicious game it is when the conker misses and hits your knuckles, and frustrating too when the chestnut you have worked so hard to get to the perfect pitch of polished hardness is shattered with a single blow.

**The palpable pride of possession.**

Being a purist at heart I have concentrated on the nut in its natural form. But nuts now come in many guises, as snacks and nibbles and all sorts of fancy – and expensive – ways of tempting people to buy them. And not only to eat. Pistachios and palm kernels, both of which are culinary rather than botanical nuts, are made into oil and butter used for cosmetics as well as cooking. (Palm oil, distinct from palm kernel oil, is made from the pulp of the fruit and is now widely being used as an ingredient in biodiesel, though with worrying environmental implications.) The macadamia, also used cosmetically, is a recent import from Australia, discovered by Europeans only in 1828 although long enjoyed by the aborigines. Difficult to extract from their hard shells, macadamias are usually sold shelled and packeted but are recommended for their dietary benefits, being rich in the 'good' fats plus a host of minerals.

The use of nuts, in their many forms and varieties, is a constant in human history from their first recorded appearance in the seventh century BC – the pistachio, in what was then Persia – to their latest contribution towards our need for fuel. But humans are not the only ones who rely on nuts.

**Vermin and nut thieves they may be, but squirrels too have to live.**

Squirrels are most closely associated with nut eating, and storing, for this is their main source of nourishment, but other rodents eat nuts too, as of course do birds in the wild. Feeding birds as a hobby is now big business, and every supermarket and garden centre will offer an array of packeted seeds, nuts and fat balls, with fancy swinging holders and tables to put them in and on. In this our nanny age we are warned to be careful because nuts can choke some birds, as they may children – whether birds also suffer from allergies I am not sure.

Mice are diligent nut gatherers, and I have several times when turning out odd corners of shed or garage – once memorably inside the bonnet of my car – come across a heap of discarded shells, hazel or chestnut, the detritus of a long winter successfully survived.

Finally, if you are a believer in old wives' tales, you might try carrying a double hazelnut in your pocket to relieve you of toothache. Without knowing this, and without knowing why, I used to hold a double almond shell in my hand on occasions that I knew were going to be stressful. It seemed to work, though who knows whether it was the nut or its doubleness that did the trick, or whether it was all in the mind.

# SEED

In a visually compelling image, Goethe described the seed as 'a germ of the future, / Peacefully locked in itself . . . / Leaf and root and bud, still void of colour, and shapeless'. Seeds are the symbol of renewal: discarded, desiccated, seemingly impenetrable, they store the makings of one of nature's most spectacular miracles.

And nature is seldom more inventive than in the ways she has devised for getting the seeds distributed. Scattered, blown, shaken and propelled; podded, burred and furred, spiked, tufted and winged; buried within the softest fruits and the hardest shells; carried in beaks, stomachs and faeces; disguised and camouflaged, hidden and brazen – the ways of seeds are multifarious indeed.

The seed itself is relatively simple. I say 'relatively' because nothing in this miraculous process is simple, but compared to the above ingenious devices the workings of the seed are straightforward once it reaches its goal: the most propitious place to germinate.

The finest seeds are usually wind borne. Each tiny poplar seed is attached to a hair, these forming the bundles of silvery-white fluff that float off the trees in late spring, settling in drifts like snow. Their fineness is proven to anyone trying to sweep them up, as they float jauntily over the broom and settle on the other side. For this in Spain they are known as 'poplar fairies'; in America the trees are called cottonwoods.

**Snow in summer: poplar seeds.**

Willows have small seeds similarly on threads but these are less acrobatic than the poplar, possibly because the willow has such a propensity to propagate by rooting that its seed distribution is not so critical.

**Prolific keys of the white ash, native of North America.**

Many conifers have wind-borne seeds with a different method of getting launched. Pine cones bearing small winged seeds, such as our native Scots pine, respond to the current weather rather than to barometric pressure, opening only when conditions are right: dry enough for the seeds to be caught on the wind but not so dry that they are unlikely to germinate. Other species of pine with bigger seeds need the help of birds for their dispersal. Some open naturally to release the seeds, but others have to be prised open by the birds. Yet others can remain locked in the cone for many years until the perfect conditions prevail, when their release is triggered by an event such as a forest fire. The large seeds we know as pine nuts are released when the cones open automatically at a certain temperature, and can be harvested by putting the cones in a sack and leaving them until they open, then hitting the sack so that the seeds fall out. They can also of course be harvested on the ground, which explains the puzzling sight in Mediterranean countries of grown men grubbing around on their hands and knees for hours at a time clutching a small handkerchief; picking up pine nuts is a very slow business.

Considering the design of wind-borne seeds inevitably leads to using words such as helicopter, parachute and glider, and it is obvious how this has come about. All these so called man-made inventions were inspired by seed flight. The commonest is the sycamore, whose ubiquity is testament to the success of these mini-helicopters that hover around our heads in late autumn. Less familiar and even more remarkable is the blade of the membrane enclosing seeds of the tree of heaven, twisted at one end so that it spins longitudinally along its axis. Ash keys are so called because they hang in bunches, often staying on the tree after the leaves are down; their eventual flight is more of a spin. Other wind-borne seeds, such as the elm and jacaranda, have a single seed within a translucent oval case which, lacking these aerobatic qualities, merely flutters to the ground – or as far as the wind happens to take it.

**The sycamore key, aerodynamic prototype.**

Weight and size are necessarily relevant to the seeds' method of dispersal, and only the lightest will be able to hitch a ride on the wind. The alder, which thrives in water, uses that as its carrier. Its female flowers form knobbly cones that release the seed in autumn – seed that has airtight cavities for buoyancy as well as a coating of oil to protect it as it floats off down the stream to find the ideal place to germinate.

Seeds are designed to wait for the right conditions before germinating. Most have a hard protective coating which must be broken down before the water and air needed for germination can penetrate. Some tree seeds – the ash for example – will achieve this by remaining dormant for up to two years after they fall, the seed cover disintegrating very slowly over this time. In others, an outside agent is needed to penetrate the seeds' defences, in the form of frost, rain, heat, the gizzards or digestive systems of animals and birds, or the biting action of insects such as termites. This scarification is imitated artificially by market gardeners and seed producers, who then go on to 'stratify' the seeds, subjecting them to the necessary low temperatures (+1° to +3°C) and moisture for anything up to three months before planting them.

The lasting properties of seeds have been the source of many wild claims, but a well-documented date palm seed, carbon dated to 2000 years old, was excavated in the palace of Herod the Great in Masada, Israel, and successfully germinated. When a seed is dormant its embryo is in a state of suspended activity, the equivalent of an animal in hibernation; that it can remain so for thousands of years is hard to encompass, though perhaps not as hard as the fact that it can then, given the right conditions, spring into life.

Since they play such a big part in pollination, it is surpising that so few insects are involved in seed dispersal. This is perhaps because the smaller seeds go with the wind, leaving only those too heavy for the average insect to hump around. The exception are those tireless workers, the ants. In a process that is mutually beneficial, the ants carry the seeds to their underground chambers, sometimes a considerable distance from

**Mimosa seeds carried by red ants.**

the host tree; many are lost on the way, and of the remainder many more will flourish in the rich soil of the nests, particularly when the ants move on to adjacent chambers, leaving a stockpile of seeds behind. This is particularly useful in arid areas as the seeds are protected both from the elements and from other predators, and put in conditions where they are most likely to germinate. In Australia, where seed dispersal by insects is the most common form, bees are drawn to some of the eucalypts for their resin which the bees need to build their nests, spreading the seeds as they come and go.

But by far the most prolific distributors of bigger seeds are birds and animals. To this end, tree seeds are dressed up in a variety of enticing packages, from cones, to berries, to nuts, to fruit.

Although berries and fruits make the pickings easy for many birds, some have to work harder for their living. Several members of the finch family – crossbills, pine grosbeaks, pine siskins – have evolved beaks that can extract pine nuts, and the agile nuthatch and some woodpeckers also feed on these nuts. (The kernels inside nuts are, of course, seeds.) Jays, and I think magpies though I've never actually caught them at it, take hazelnuts, and some bird – I never discovered which – holed and dropped far too many walnuts on my farm in

**Coal tits feed on pine nuts, as do many members of the finch family.**

**The spindleberry tree is inconspicuous until these fruits burst out.**

Spain. This appears to be a one-sided affair with only the birds winning out, though I suppose the odd nut gets carried further from home than if it had just dropped off the tree, and is left intact to germinate in its new site.

What we term loosely as berries seem to have been made for birds. Their colour, their size, their accessibility and texture must be irresistible seen through a bird's eye, a supposition borne out by the way they gobble them up. Think of the glowing, seductive colours of the berries of hawthorn (the 'haws' of hips and haws), rowan and elder. But this is far from one-sided, as not only do the seeds get dispersed by the birds, they are also activated in their passage through the birds' digestive system. Some mammals, such as the pine marten, provide this service too, but squirrels are mere hoarders and only chance or a slip of the memory allow for some of their booty to survive and flourish away from the parent tree.

The bright orange seeds of the spindleberry nestle inside a four-lobed magenta casing, both colour and shape surely inspiration for a milliner. Birds find the seeds highly attractive, but they are best given a wide berth as they are poisonous to humans. The seeds of the wayfaring tree also seduce us with colour, their small conical berries in clusters of green, yellow, red and black, changing as they ripen. They too are mildly toxic, though not to birds.

Talking of birds brings to mind a plant that depends on both birds and trees for its existence – the mistletoe. Inevitably associated with Christmas and kissing, and familiar to anyone living in rural France, mistletoe is described as hemi-parasitic, that is it draws water and minerals from the bark of the host tree but does its own job of photosynthesis. In Europe mistletoe grows mainly on poplar and apple trees, and in quantities that can seem to overwhelm its host. The seeds of mistletoe are spread by birds, which eat the fruits and then wipe the sticky pulp off their beaks on to a tree branch, leaving the seeds embedded in the bark. (They also get passed through their droppings.) The Latin name for European mistletoe is *Viscum album*, meaning 'sticky white', and it gives its common name to the mistlethrush, a speckled thrush with a big belly that looks as if it gets plenty of nourishment from the berries.

Although mistletoe is usually perceived as a pest that may deplete and even kill its host, it has many positive attributes, such as the 'witches' brooms' provided by the tangled branches of different mistletoes in Australia, where an extraordinary 75 per cent of birds – about 240 species – are recorded as nesting in them. And although the berries are poisonous, they are widely used medicinally. Linking the parasitic nature of the plant to a cancer has led to the leaves and berries of mistletoe being used, controversially, as a natural anti-cancer treatment,* and under the name of Viscum album it is a homeopathic remedy for rheumatic and gouty complaints as well as – fittingly, given its position in the treetops – for vertigo.

The terminology for seeds, as with many other parts of the tree, is bewildering to the non-expert. Words such as 'pea' and 'bean' are used indiscriminately to describe any large seeds resembling them, and often apply to the pod as well as the seed within.

The Judas tree's spectacular flowers turn into flat, woody seed pods from which the tree takes its generic name, *Cercis*, Greek for shuttle, since the pods resemble a weaver's shuttle. They come in beautiful shades of purple and brown, and the seeds are like small dried peas appropriate to the flowers.

**This dazzling array of eucalyptus pods was photographed in Porto.**

Of all the seed pods, some of the most ingenious are those of the eucalyptus, looking like fired clay artefacts designed by some elfin master-potter. 'Pod' is not the right word; botanically it is a capsule, but I would prefer to call it a casket – not as the Americans understand that word, but in the sense of a carefully made container for something precious, a jewel box for the tiny seeds. These containers are the fruit of the tree, and it takes its name from them; eucalyptus is the Greek for 'well covered'.

Another very hard seed, and a true seed despite its name, is nutmeg. Indigenous to the so called Spice Islands of Indonesia, it is now grown there and in the Caribbean as well as in Kerala, in southern India. It is an evergreen with glossy leaves, related to the magnolias, and produces not only the nutmegs we are familiar with, that keep for ever and rattle so appealingly in their glass jar – unless, shame on you, you buy them already grated – but also a less common spice, mace. Though in the West nutmeg is probably most used for baking, in India it is valued as a spice in curries. An essential oil obtained by distilling ground nutmeg plays a part in both medicine and cosmetics.

I would like to say that coffee as we know it is made from the seeds of the coffea tree, but this is making things far too simple. First, they are always known as beans, which they are not. Second, most descriptions tell you that coffee beans are berries, though some call them cherries; in fact they are the seeds from inside the berries. Lastly, they are often said to be produced on small bushes or shrubs, which they may be for the purposes of harvesting the beans but if left to their own devices these 'bushes' can grow to a height of twelve metres, surely a tree by any standards.

And to continue in this vein, the Indian bean tree or catalpa does not come from India, does not bear beans, and should be called not catalpa but catawpa, the native American tribe after which it was wrongly named. Poor bean tree – but with its startlingly beautiful scented flowers, heart-shaped leaves, majestic bearing and distinctive dangling pods (which contain papery winged seeds), does anyone really mind what it is called? It even secretes nectar from its leaves, most unusually. The catalpa has been a favourite tree for parks and public gardens since it was introduced from the southern states of America in 1762. 'In full flower in late summer it is one of the best trees we possess,' writes Roland Randall in *Trees in Britain*.

As influential as coffee, and classified as a tree though never growing to more than about eight metres, is the cacao, whose seeds, known as beans, nestle in a white pulp inside the fruit, which is called a pod. Confused? Never mind, just enjoy the end product. Full of fat (up to 50 per cent) and containing an alkaloid similar to caffeine, the seeds produce that most wicked of substances, chocolate. The history of cocoa goes back 4000 years to the Amazon basin, but it was not until the sixth century AD that the Mayans discovered the delights of chocolate, which they took mainly as a thick unsweetened drink they called *xocolatl*, 'bitter water'. By 1200 the Aztecs had become addicted; three hundred years later the emperor Montezuma II allegedly drank up to fifty goblets a day, and soon beans were being used as currency. It was Cortés who brought cocoa beans to Europe in 1528 as a gift for Charles V, and the Spanish who turned them into something like the chocolate we know today – a secret they managed to keep for a hundred years.*

Chocolate is so engrained in our culture that it is hard to associate it with its origins: an evergreen tree whose clusters of pale delicate flowers, like those of the Judas tree, grow mainly out of the trunk as well as on some branches. These small flowers turn into ovoid lanterns which mature from green to yellow and red, when ripe almost purple. From this profusion of flowers, 6000 or so per tree, only about twenty will set into 'fruit' – more would surely break the tree, for these tough leathery pods will measure up to 30 centimetres long and weigh as much as half a kilo. Each contains 20 to 50 of the precious beans, and it takes about 500 to make a kilo of chocolate. And if by chance there are any left after man has harvested them, they will be dispersed by monkeys, who take the fruit for its pulp and discard the bitter seeds.

Tempting though it is to pursue the seeds of the cacao tree further on their very different cycle from the Amazon through the chocolate factory into the world's shops and back into the earth, we must return to the damp, green woodland of Britain, where last year's seeds are beginning to stir and stretch as they prepare for that prodigious leap from dormancy to activity that will result in one more season of growth, of leaf, flower and fruit, as one more ring is added to the tree's trunk.

**Dazzling too are the colours of cacao pods, changing as they ripen.**

EPILOGUE

# *Trees for the future*

What of the future for trees? And is this a question even to be asked for something that has lasted 400 million years and could outlive us humans by as many more?

As I am not a conservationist, nor is it that sort of book, I have no answers to these questions, whether they are valid or not. Taking the close-up, personal view I think all we can do is to maintain our sense of awe and to respect trees wherever we find them, and to pass on this awe and respect to our children so that they may in turn pass it on. And perhaps plant the odd fruit tree or acorn now and then, both imaginatively and in reality.

There's a delightful parable told by the French author Jean Giono called 'The Man Who Planted Trees', about a shepherd in the Var department of Provence who does just that – on his own and persistently, over forty years, transforming the barren hillsides into lush woodland and bringing water and new life to the local villages. The story has a twist in its tail which I won't divulge, as it's a story worth reading.

Planting trees is one of the most deeply satisfying things I know, and I am not alone. 'The true meaning of life is to plant trees, under whose shade you do not expect to sit.'* This reminds me of how outraged I felt when someone who had been commissioned by a friend to buy me a tree turned up with a potted plant instead, saying 'What's the point in planting a tree when you won't be around to see it grow?' Fortunately such an attitude is rare, and the symbolic act of planting trees continues. In New York a 'Tree of Hope' was planted in November 2003 in the churchyard of St Paul''s Chapel near the World Trade Center: a Norway spruce, replacing a sycamore destroyed by the fall of the towers two years earlier. The last living object to be rescued from the

**Symbol of resilience and hope: the wounded Callery pear** (above) **blossoms once more in its new incarnation as the Survivor Tree on the World Trade Center Memorial Plaza.**

rubble had been a pear tree, little more than a blackened torso; after nine years of nurturing, it was ceremoniously replanted in the Memorial Plaza. At the same time began the task of planting four hundred swamp white oaks that will grow to 25 metres and provide a woodland oasis on this much visited site, a visual reminder of the cycle of life and death.

If men and women have lived in the woods – and in medieval England entire woodland communities occupied cottages and farmsteads scattered throughout the woods, connected by a network of cart tracks and footpaths – they must also have died and been buried in them, until Christianity tidied things up, in the way it likes to do. But recently people have reclaimed their right to be buried, or scattered, as they please, and one of the most popular alternatives to a church service which few believe in or the almost farcical performance of the cremation ceremony, is woodland burial. This makes sense in so many ways. It is ecologically sound, since the body is placed in a coffin made of willow or wicker or papier-mâché, all rapidly biodegradable, or even a woollen shroud, and it – or the ashes – are marked not by a headstone but by a small plaque laid flat on the ground beside the accompanying tree. This tree can be chosen from a list, all native species (just think of providing sustenance to a tree), and will form part of an eventual woodland, to be protected – unlike graveyards – in perpetuity.

Trees have a gift for regeneration. The yew, associated in many cultures with death and renewal, in Breton legend is depicted with its roots in its mouth: the tree's equivalent of the serpent eating its own tail. The irrepressibility of tree growth is exploited in the old practices of coppicing and pollarding, of which our garden pruning is a descendant. It's a useful allegory for humans too: cutting back exuberant growth and dead wood in an act of love and skill as opposed to wilful – or mindless – destruction, will bring high rewards for both the cutter and the cut.

As we have seen, nature drums up many ruses for helping trees survive: seduction, camouflage, dormancy, symbiotic relationships and many more. And her timescales are not ours. Eight thousand years ago large parts of the Sahara were covered in trees; it is estimated by those who study the climatic cycles that in another fifteen thousand years, that is in AD17,000, it will be green once again.

Meanwhile, man has to make some changes. The supplies of fossil fuels laid down over the millennia, much of which come from trees, are not infinite though our demand for them appears to be. Consciousness of the need to look after our trees is creeping in slowly. In a small gesture from China, whose forests can't keep up with the demand for 80 billion throw-away pairs of wooden chopsticks a year, the Chinese are turning to plastic. (Though how that exchange pans out in terms of ecological gain is not made clear.) As we have seen, palm oil is a valuable source of biodiesel, though with equal reservations about the environmental effects. More promising is the use of willow as a biofuel, now gaining some ground. The willow is reduced to chips which when burnt give off a strong heat, producing no more, possibly less, carbon dioxide than they absorbed while growing. Some people are growing, cutting and drying it themselves – and willow grows very quickly.

One of the most exciting ideas for generating alternative power, and the most sympathetic to our subject, is the recent proposal for wind trees. Still in the design process, these are based on a simplified tree form of trunk and leaves. Each of the hundred or so leaves works as a tiny windmill, much more sensitive to the wind than any conventional windmill and needing less than half the wind power: 2kph as opposed to the 5kph needed to power the huge white birds now soaring – controversially – across much of the countryside of Europe. At heights of 8 or 12 metres, the proposed wind trees can be bought and used on a domestic scale, one or two per house, providing an individual power supply and, I think most

**These trees of the future will combat pollution in novel ways – here by providing plug-in points for recharging electrically powered vehicles.**

people would agree, adding aesthetically to the urban environment. Surely this is a significant event in the long history of trees.

Awareness gained in front of the television screen or at the computer is valuable, and nature programmes bring the wonders of trees into our sitting room, including rare and exotic ones we are never likely to see. But there is no substitute for the real thing. Whether in groups

and organisations – and there are many – or on one's own, in garden or park or local wood, being among trees is restorative. Their stability in a shifting world, their age, their character and their demonstration of the life-cycle that we share with them, are all food for the soul. Every single tree is unique, and whatever our perceptions or prejudices, not one of us, foolish or wise, sees the same tree.

# Notes

Notes providing additional information are marked in the text with an asterisk; straightforward references are unmarked.

**p.6** The black poplar, subspecies *Populus nigra betulifolia*, is one of Britain's rarest native trees. This fine example by the river Stort near Roydon in Hertfordshire is one of only about 600 females out of an estimated total of 10,000 black poplars. Alan Burgess painted all fifty of the Great British Trees selected by the Tree Council in 2002 to mark the Queen's Golden Jubilee.

**p.7** William Blake (1757-1827), one of the Proverbs of Hell from *The Marriage of Heaven and Hell*, 1790-93.

Joyce Kilmer (1886-1918),'Trees', 1913.

John Muir (1838-1914), *Our National Parks*, Houghton, Mifflin, 1901.

**p.9** Oldest wooden artefact: Claims for this honour vary wildly, but the Clacton spear discovered in 1911 is reliably dated by the Natural History Museum as 450,000 years old. Eight spears (of what wood is not stated) found in Schöningen, Germany, are estimated to be between 270,000 and 400,000 years old.

Evolution of trees: During the Devonian period, *c*.419-359 million years ago, trees evolved from primitive ankle-high shrubs to trees up to 30 metres tall. A perfectly preserved fossil tree of 8 metres, dated to 385 million years ago, was discovered at Gilboa, New York, in 2007.

**p.13** Sahara: Roland Oliver, *The African Experience: From Olduvai Gorge to the 21st Century*, Phoenix Press, London, revised ed, 1999, p.39.

*Epic of Gilgamesh*: the exploits and travels of Gilgamesh, king of the ancient Sumerian city Uruk in the twenty-seventh century BC, written in cuneiform on clay tablets.

**p.15** Gordion: capital of the ancient kingdom of Phrygia, 80 kilometres southwest of Ankara.

**p.17** *Rhamnus cathartica*, from *A Modern Herbal*, Maud Grieve, 1931.

**p.19** Alfred, Lord Tennyson, 'The Oak', 1889.

'Heart of Oak', original words by David Garrick 1759 (later amended), music by William Boyce. In Garrick's version the lyric went 'Heart of oak are our ships, jolly tars are our men.'

W.H. Auden, 'Woods', 1953.

**p.20** Sweet chestnut photographed at Chirk Castle, Wrexham.

**p.22** 'One learns more in the woods than in books; the trees and the rocks will teach you things that you would not otherwise be able to understand,' wrote Saint Bernard of Clairvaux, founder in 1118 of the Abbey of Fontenay in the then heavily wooded landscape of Burgundy.

**p.23** Giuseppe Arcimboldi, or Arcimboldo (1537-93).

**p.24** Henry Moore: *Reclining Figure* 1935-36 was the first of six elmwood figures Moore carved; he bought the trunk from a wood merchant in Canterbury, and learnt much about the properties of wood as the elm dried and moved. The studies of Dorothy Hodgkin's hands came about as a request from the Royal Society for a portrait of their distinguished Fellow, who in 1964 had been awarded the Nobel Prize for her work in X-ray crystallography, carried out despite her crippling arthritis.

Michael Viney, *A Year's Turning*, The Blackstaff Press, 1996; Penguin Books, 1998.

**p.25** Kahlil Gibran (1883-1931) thought and wrote about nature in a way that was well before his time, inspired by the countryside of his native Lebanon.

**p.26** David Hockney: Interview in the *Guardian*, 27 March 2009. The corrected version appeared in the *Daily Mail*, 28 March 2009.

**p.28** William Fairbank, *The Forest Stations*, Frontier Publishing, 1998.

Though the Orpheus poem is usually attributed to Shakespeare it may have been by John Fletcher.

**p.30** Lombardy poplar: 'Italica', a cultivar in the varied family of the black poplar, *Populus nigra* – very different from the stumpy, leaning and often warty trunks of the subspecies *P. nigra betulifolia*, see illustration on page 6.

**p.36** Cinchona bark: 'The discovery of homeopathy started here. Hahnemann had given up the practice of medicine due to his observations that it often caused more harm than good. He took to translating medical texts and came across the account of cinchona bark which was said to be a remedy for malaria due to its bitter properties. He decided to take some and subsequently developed remittent fevers and other symptoms similar to those of malaria. This gave him the idea of the law of similars.' My thanks to Lucinda Torabi for this account.

**p.43** The driftwood bird, most probably formed by an alder root, was found in this exact state (though bleached white) on the side of Loch Mullardoch in the Northwest Highlands of Scotland in 1951, when the river Cannich was dammed to turn the loch into a reservoir. Sculptures formed by tree roots and stumps from the old Caledonian forest that once covered the Highlands, preserved in peat over thousands of years and washed into the lochs, are now big business.

**p.44** Love of money: Timothy 6, 10.

**p.47** Alex Haley: *Roots: The Saga of an American Family* was published in 1976 and made into a TV mini-series the following year; the book was translated into 37 languages and won a special Pulitzer Prize.

Marguerite Wilkinson, 'A Woman's Beloved', 1917.

Dylan Thomas, 'The Force that Through the Green Fuse Drives the Flower', 1934.

**p.49** The word inscape was coined by Gerard Manley Hopkins to mean those features which make every landscape or natural structure unique; 'instress' was how he described our reaction to inscape.

**p.51** Paul Blissett, a professional hedgelayer, kindly donated this photograph of a hedge laid by him and a fellow craftsman. Paul's website is www.hedgelayer.freeserve.co.uk.

**p.52** Both quotes are from Godfrey Baseley, *A Country Compendium*, Sidgwick and Jackson, 1977.

**p.54** Adam Foulds, *The Quickening Maze*, Jonathan Cape 2009, p.59. Reprinted by permission of The Random House Group Limited.

**p.61** Philip Larkin, 'The Trees', 1974.

**p.62** Robert Frost, 'Birches', 1915.

**p.65** 'The neem tree could have been designed by a celestial committee (maybe it was). A collaboration of genetic engineers, chemical engineers, pharmacists, agronomists, and dieticians could not have produced a more interesting, and some say, valuable, plant.' Steve Nix, About.com Forestry.

**p.69** Michael Viney, *A Year's Turning*, The Blackstaff Press, 1996; Penguin Books, 1998, p.196. Quoted by kind permission of the author.

**p.71** Charles Tomlinson, 'The Beech', 1969, *New Collected Poems*, Carcanet Press, 2009.

**p.73** 'Trials carried out in the US and Sweden have shown that when people move into a green, leafy environment, symptoms of stress can be measurably reduced within four minutes. Pulse rates and perspiration are lowered, the muscles across the forehead relax and there is generally an increased sense of wellbeing.' *Broadleaf*, the Woodland Trust magazine, spring 2003. My thanks to Paul Blissett for pointing this out.

The story of Little Red Riding Hood has many different endings, and even the Brothers Grimm toned down this grisly one in later versions.

**p.74** Caspar David Friedrich (1774-1840); see for example *The 'Chasseur' in the Forest*, 1813.

**p.75** Gilbert White, *The Natural History of Selborne* (first published 1788-89), edited by Richard Mabey, Penguin Books, 1987, p.30.

Composers: In Vivaldi's *Four Seasons*, the goatherd sleeps under the rustling branches, while nymphs and shepherds 'dance beneath the brilliant canopy of spring'. Beethoven escaped from Vienna and the problems of his domestic life and encroaching deafness to walk in the country, and the first movement of the Pastoral Symphony is annotated: 'Awakening of cheerful feelings upon arrival in the country'. And the one glimmer of light in Schubert's gloomy song cycle *Winterreise*, Winter Journey, comes in the fifth song, 'Der Lindenbaum', when the lime tree offers comfort to the grief-stricken lover.

Henry David Thoreau, *Walden, or Life in the Woods*, 1854.

**p.76** John Fowles, *The Tree*, Aurum Press, 1979, n.p.

**p.77** Ted Hughes, 'A Tree', 1979.

**p.82** D.H. Lawrence, 'Almond Blossom', written in Fontana Vecchia, Taormina, Sicily, 1920/21.

**p.87** William E.Vaughan (1915-77) was an American journalist known for his aphorisms.

**p.89** Nutritional supplements: This huge industry makes wide use of tree products, from the standard vitamin C that comes from citrus fruits to others with various claims extracted from ginkgo and hawthorn leaves, the bark of maritime pine, green (unroasted) coffee beans, sweet chestnuts, olives and pomegranates.

**p.91** Ted Hughes, 'Fulbright Scholars', *Birthday Letters*, Faber and Faber, 1998.

Andrew Marvell, 'The Garden', 1650.

**p.93** 'Lemon tree . . .': folk song written by Will Holt, 1960s, based on an earlier Brazilian folk song.

**p.94** There are many versions of the *Vitae Adae et Evae*, a Jewish work from the first century AD also translated into Latin, Greek and Armenian.

**p.95** John Fowles, *The Tree*, Aurum Press, 1979, n.p.

**p.99** Fig: Wikipedia, Common fig.

D.H. Lawrence, 'Bare Fig Trees', Taormina, 1920/21.

**p.101** Richard Mabey, *Food for Free*, HarperCollins Publishers, p.38. © Richard Mabey 1972, 1989, 2007, 2012. Reproduced by permission of Sheil Land Associates. Page numbers cited refer to the 1989 edition.

**p.102** Bananas grow not on trees but on herbaceous flowering plants, the genus *Musa,* which includes the plantains. The plants grow from a corm; the stem of the plant is formed from the base of the leaves and can reach as tall as 7 metres. The bananas grow out of a single flower spike in tiers surrounding it, giving the familiar 'hands' that conveniently hold the fruit together.

**p.103** Mabey, *Food for Free,* pp.44-5.

**p.107** Chestnut sellers: Godefroy Durand, from *L'Illustration: Journal Universel,* Paris, 1857.

**p.108** Mabey, *Food for Free,* p.30.

**p.113** Goethe, from 'The Metamorphosis of Plants', 1797.

**p.116** The spindleberry photograph was kindly supplied by Simon Taylor, who runs Little Trewern B&B in the Black Mountains of Wales.

**p.118** Mistletoe: much of the controversy centred on the work of the Swiss educational philosopher Rudolf Steiner, who in the 1920s pioneered the use of mistletoe extract in treating cancer. Under the trade name of Iscador, mistletoe was used by the American actress Suzanne Somers in a much publicised crusade for natural alternatives to chemotherapy in treating her own breast cancer.

**p.120** Roland E. Randall, *Trees in Britain, Broadleaved Book 3,* Jarrold and Sons Ltd, Norwich, 1975, p.31.

The story of chocolate makes sobering reading. In the early sixteenth century a slave could be bought for a hundred cocoa beans. Despite the Abolition Act of 1807, slave labour continued to be used on cocoa plantations in central Africa well into the twentieth century, and child labour continues today. In a TV documentary made in the early 2000s, a plantation worker was asked what he would say to consumers of chocolate, something he himself had never seen. He replied, 'They are eating my flesh.' See Catherine Higgs, *Chocolate Islands: Cocoa, Slavery and Colonial Africa,* Ohio, 2012.

**p.123** Jean Giono, *The Man Who Planted Trees,* The Harvill Press, 1996.

'The true meaning of life . . .': Nelson Henderson, an Irishman who emigrated to Canada at some time in the nineteenth century, is known for little else than this well-worn but still worth quoting sentiment.

**p.125** L'Arbre à Vent, New Wind R&D; for more details see www.arbre-a-vent.fr.

**pp.126-7** Old mixed woodland – cork and holm oak, chestnut, pine and olive – above Galaroza in the Sierra de Aracena.

# Index

The index is selective.
Numbers in italic refer to illustrations.

# Picture credits

Antonio Abrignani/Shutterstock p.107 (right); Barbara Aldiss p.119; Paul Aniszewski/Shutterstock p.8; Subbotina Anna/Shutterstock p.104; arbopals.com p.114; Pierre-Yves Babelon/Shutterstock p.121; bitemywords.com p.107 (left); Paul Blissett p.51; Alan Burgess p.6; cafeclock.com p.97; John A. Cameron/ Shutterstock p.16; Bharati Chaudhuri/SuperStock p.45; © Christie's Images/SuperStock p.68; Thomas Curtis p.10; dabjola/Shutterstock p.85; Simon Deans (deansfamily.com) p.116; diak/Shutterstock p.57; Scott Duncan/Flickr p.12; Angela Dyer pp.34, 41, 82, 92, 99, 102, 126-7; Flaviano Fabrizi/Shutterstock p.84; FactZoo.com p.63 (right); William Fairbank p.27; g215/Shutterstock p.66; Rory Francis/Flickr p.20; Eric Gevaert/Alamy p.65; Henry Moore Foundation, reproduced by permission p.24; hodnet.org.uk p.95 (top); Tony Howell pp.22, 46, 70; Pavel Ilyukhin/Shutterstock p.53; Aaron Johnson © 2006 p.59; Robert Kawka p.109; Julie Bloss Kelsey p.39; Derek Langley/darknessandlight.co.uk © 2013 p.31; 1.bp.blogspot.com p.63 (left); Mark Lennihan/AP/PA Images p.122; leslieland.com p.89; life_in_a_pixel/ Shutterstock p.58; loveoneanother.us p.18; Jaroslav Machacek/Shutterstock p.88; Alexandr Makarov/ Shutterstock p.67 (left); Uroš Medved/Shutterstock p.106; David Orcea/Shutterstock p.90; Pefkos/ Shutterstock p.60; Vadim Petrakov/Shutterstock p.48; Salim October/Shutterstock p.33; David Russell p.43; SeDmi/Shutterstock p.67 (right); Simon Taylor p.117; David Thyberg/Shutterstock p.38; treepicturesonline.com p.95 (bottom); vermilionriver.blogs p.112; Christian Vinces/Shutterstock p.14; Zdorov Kirill Vladimirovich/Shutterstock p.110; WDG Photo/Shutterstock p.40; Ivonne Wierink/ Shutterstock p.54; © Jonty Wilde/David Nash, courtesy of YSP p.23; wisarts.com p.113; WTPL/Jane Corey p.72; www.arbre-a-vent.fr p.125; www.classic.co.uk p.77; www.nhm.ac.uk p.87; www.prabhupada.es p.115; Jess Yu/Shutterstock pp.2-3.
Every effort has been made to trace copyright holders; any omitted or incorrectly credited should contact the publisher.